MICROELECTRONIC SYSTEMS

MICROELECTRONIC SYSTEMS
Design, Modelling and Testing

Dr W Buchanan

Senior Lecturer, Department of Electrical, Electronic & Computer Engineering,
Napier University, Edinburgh

A member of the Hodder Headline Group
LONDON • SYDNEY • AUCKLAND

Copublished in North, Central and South America by
John Wiley & Sons, Inc., New York • Toronto

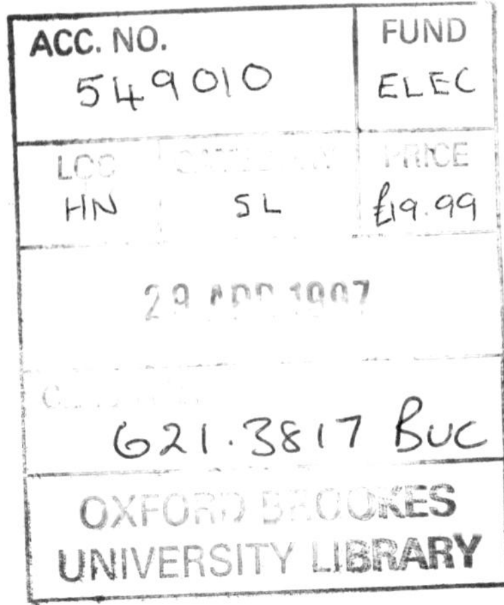

First published in Great Britain in 1997 by
Arnold, a member of the Hodder Headline Group,
338 Euston Road, London NW1 3BH

Copublished in North, Central and South America by
John Wiley & Sons, Inc., 605 Third Avenue,
New York, NY 10158-0012

British Library Cataloguing in Publication Data
A catalogue record for this book is available from the British Library

Library of Congress Cataloging-in-Publication Data
A catalog record for this book is available from the Library of Congress

ISBN 0 340 67771 6

Printed and bound in Great Britain by JW Arrowsmith, Bristol

This book is dedicated to my darling wife, Julie, for the help and support she has given me over the years.

Contents

Contents vii

Preface xiii

1. Introduction 1

1.1 History of microelectronics 1
1.2 Electronic design 2
1.3 System development 3
1.4 Design abstraction 5
1.5 IC design process 7
1.6 Logic technologies 7
1.7 Jargon 15
1.8 Exercises 16
1.9 Postscript: From the Bell Laboratories to Microsoft 17

Part 1: Design

2. Bi-polar Design 21

2.1 Introduction 21
2.2 TTL design 21
2.3 I^2L design 30
2.4 ECL design 31
2.5 Exercises 36

3. MOS Design 39

3.1 Introduction 39
3.2 MOS devices 39
3.3 CMOS logic gates 40
3.4 Designing CMOS logic 42
3.5 PMOS logic gates 45

3.6 NMOS logic gates 46
3.7 CMOS .vs. TTL 47
3.8 Noise immunity 47
3.9 Bi-CMOS 47
3.10 MOS precautions 48
3.11 Exercises 49

4. MOS Device Modelling 51

4.1 Introduction 51
4.2 MOS characteristics 53
4.3 Gate capacitance 53
4.4 MOS devices as a switch 54
4.5 NMOS inverter with a depletion load 56
4.6 NMOS NAND and NOR gates 58
4.7 Exercises 60

5. NMOS Circuit Design 63

5.1 Introduction 63
5.2 NMOS fabrication process 63
5.3 Stick diagrams 64
5.4 Lambda design rules 66
5.5 Contact cuts 68
5.6 Depletion-mode load transistor 68
5.7 NMOS logic circuit design 69
5.8 Electrical characteristics 73
5.9 Comparison of NAND and NOR gates 75
5.10 Exercises 75

6. CMOS Circuit Design 77

6.1 Introduction 77
6.2 CMOS processing 77
6.3 CMOS inverter 78
6.4 CMOS NAND gate 79
6.5 CMOS NOR gate 81
6.6 Electrical characteristics 82
6.7 Exercises 82

7. Practical Design of an ASIC 83

7.1 Introduction 83
7.2 Requirements analysis 83
7.3 Requirements specification 85
7.4 Design 88
7.5 Simulation 89
7.6 Implementation 90
7.7 Exercises 98
7.8 Projects 100

8. **Memory Systems** **103**

8.1 Introduction 103
8.2 Memory addressing size 104
8.3 ROM 105
8.4 2D addressing 106
8.5 Static RAM (SRAM) memory 108
8.6 Dynamic RAM (DRAM) memory 108
8.7 DRAM memory cell 111
8.8 Sense amplifier 112
8.9 Memory cache 114
8.10 Exercises 115

9. **Field Programmable Gate Arrays** **117**

9.1 Introduction 117
9.2 Programmable devices 117
9.3 FPGA classifications 118
9.4 FPGA logic blocks 120
9.5 FPGA programming types 122
9.6 Practical FPGAs 126
9.7 Conclusions 134
9.8 Exercises 134

Part 2: Modelling

10. **High-Level Modelling** **139**

10.1 Introduction 139
10.2 High-level language modelling and emulation 139
10.3 Modelling combinational logic 140
10.4 Boolean sequences 145
10.5 Sequential logic 146
10.6 Bit fields 148
10.7 File output 153
10.8 Exercises 155

11. **VHDL Modelling** **161**

11.1 Introduction 161
11.2 History of VHDL 161
11.3 Structural decomposition 161
11.4 Terminology 163
11.5 Entities 163
11.6 Architectural bodies 165

11.7 VHDL operators 165
11.8 Identifiers 166
11.9 Reserved words 166
11.10 Numbers, characters and strings 167
11.11 Constants and variables 168
11.12 Scalar types 169
11.13 Flow control 174
11.14 Exercises 180

12. PSpice Modelling 183

12.1 Introduction 183
12.2 DC analysis 183
12.3 PSpice simulation 185
12.4 Units 185
12.5 Device references and model parameters 187
12.6 Voltage sources and stimulus 188
12.7 Current sources and stimulus 191
12.8 Dot commands 191
12.9 Active components 195
12.10 Example circuits 196
12.11 Exercises 204

13. Fault Modelling 205

13.1 Introduction 205
13.2 Fault coverage and fault models 206
13.3 Path sensitization 207
13.4 Simulating stuck-at faults 211
13.5 Testing stuck-on/off faults in transistors 213
13.6 Exercises 215

Part 3: Testing

14. Designing for Test 221

14.1 Introduction 221
14.2 Combinational and sequential circuits 221
14.3 Scan-in scan-out design 221
14.4 Automatic test circuits 225
14.5 Exercises 232

15. Memory Tests 233

15.1 Introduction 233

	15.2	Test patterns	233
	15.3	Memory test program for a PC	238
	15.4	Exercises	239
A.	**MOS Characteristics**		**241**
	A.1	MOS equations	241
	A.2	NMOS invertor with depletion load	243
	A.3	Exercises	244
	A.4	Program to determine gate capacitance	245
B.	**IC Fabrication**		**247**
	B.1	Introduction	247
	B.2	NMOS fabrication	247
	B.3	CMOS fabrication	249
C.	**CMOS/NMOS DESIGN RULES**		**253**
	C.1	Introduction	253
	C.2	1.2 micron single polysilicon double metal n-well and p-well design rules and process specifications	253
	C.3	2.0 micron double polysilicon double metal n-well CMOS process specification	254
	C.4	CIF format	255
D.	**TRAFFIC LIGHT CONTROLLER SIMULATOR**		**259**
	D.1	Introduction	259
	D.2	Program and test results	259
E.	**VHDL REFERENCE**		**265**
	E.1	Conventions	265
	E.2	Syntax	265
F.	**PSpice Tutorial**		**267**
	F.1	Introduction	267
	F.2	Simulation process	268
	F.3	Schematic capture	269
	F.4	Saving file	274
	F.5	Electrical checking	275
	F.6	Specifying an analysis	275
	F.7	Netlist	276
	F.8	Simulation	277
	F.9	Plotting graphs	278
	F.10	Viewing text output	279
	F.11	Exercises	281

G. PSpice Reference **285**

G.1 Introduction 285
G.2 GaAsFET (B) 285
G.3 Capacitor (C) 286
G.4 Diode (D) 287
G.5 Voltage-controlled voltage source (E) 288
G.6 Current-controlled current source (F) 289
G.7 Voltage-controlled current source (G) 289
G.8 Current-controlled voltage source (H) 290
G.9 Independent current source and stimulus (I) 291
G.10 Junction FET (J) 291
G.11 Inductor-coupling (transformer core) (K) 292
G.12 Inductor (L) 293
G.13 MOSFET (M) 294
G.14 Digital input (N) 295
G.15 Digital output (O) 295
G.16 Bi-polar transistor (Q) 295
G.17 Resistor (R) 297
G.18 Voltage-controlled switch (S) 298
G.19 Transmission line (T) 299
G.20 Digital stimuli and primitives (U) 299
G.21 Independent voltage source and stimulus (V) 299
G.22 Current-controlled switch (W) 300
G.23 Sub-circuit Call (X) 301

H. Further Reference **303**

H.1 WWW reference 303
H.2 30-pin SIMM 303
H.3 72-pin SIMM 303

Index **305**

Selected Solutions **313**

Preface

I have taught electronic design for a number of years and one thing I have noticed is that there are very few electronic modules which try to integrate the main product development areas, that is, modelling, design and test. A properly designed system should first be modelled before a full design can take place. Design should also contain extra elements which make it easier to test and should also be modelled for faults. It is thus important for electronic engineers to have skills in all of these areas. As electronic systems are becoming more complex the need for a properly designed, modelled and tested system is extremely important. Many companies have a short development time from the initial concept to the final product, any failure in a part of the development can be costly in the long term.

Software engineers are taught the concepts of requirements analysis, requirements specification, design, implementation and test. Electronic engineers are often taught these areas in isolation to the others. This book hopefully addresses this by covering most of these areas.

The book splits into three main parts and a number of appendices. The main parts are design, modelling and test. Each area uses typical techniques and technologies in each area. The key applications areas, include:

- CMOS/NMOS/Bi-polar design principles;
- Practical example of a CMOS ASIC;
- Practical ASIC simulation;
- FPGA design;
- Industry standard modelling techniques, such as Spice and VHDL;
- Fault modelling;
- Designing for test;
- Test methodologies;
- Design of memory system;
- Fault modelling of memory systems.

Electronic engineers must now be multi-disciplined and have skills in hardware and software development. Often software modelling techniques are used for system modelling, these in the use of high-level languages and VHDL. This book contains software programs, in ANSI-C, which can be used to implement important concepts, to model digital systems and to generate test methodologies.

The appendices are used to provide a basic source of reference material. A PSpice tutorial is also included as an introduction to Spice modelling. Further information on related subjects, such as FPGAs, GaAs design, NMOS and CMOS design, as well as up-to-date modelling software is available on the WWW page:

```
http://www.eece.napier.ac.uk/~bill_b/mbook.html
```

Help from myself can be sought using the email address:

```
w.buchanan@napier.ac.uk
```

or, if unavailable, send an email to:

```
bill_b@miranda.eece.napier.ac.uk
```

Dr William J Buchanan
Napier University

Introduction

1

1.1 History of microelectronics

The science of electronics began in 1895 when Lorentz postulated the existence of charges called electrons. By 1897 Braun had built the first electron valve, which was a simple cathode-ray tube. Then at the beginning of the century, Fleming invented a diode which he called a valve. This device used a heated cylindrical plate in a vacuum. A positively charged heater plate caused a current to flow, but when negatively charged no current flowed.

In 1907, Lee De Forest made a triode by adding a grid so that a small control voltage could control a large current. Its main advantage was that it could amplify electrical signals. By the 1940s several scientists at the Bell Laboratories were investigating materials called semiconductors. These substances, such as silicon and germanium, conducted electricity only moderately well, but, when doped with impurities their resistance changed. From this work they made a crystal called a diode which worked like a valve. The diode had many advantages over the valve, including the fact that it did not require a vacuum and was much smaller. It also worked well at room temperature, required little current and had no warm-up time. This was the start of microelectronics.

In 1948, William Shockley at the Bell Lab produced a transistor that could act as an amplifier. This was a germanium crystal with a thin p-type section sandwiched between two n-type materials and it could operate like a triode. Shockley received a Nobel Prize in 1956 for this work.

By 1953, transistors were small enough to fit into hearing aids that fitted into the ear. Transistors soon operated over higher frequencies and within larger temperature ranges. Eventually they became so small that many were placed on a single piece of silicon. These are typically known as microchips and they started the microelectronics industry.

By the 1960s several transistors were put on a single integrated circuit (IC). This type of device was defined as small-scale integration (SSI). The definition of SSI is normally limited to less than 100 transistors on a single IC. By 1966, the technology had evolved to medium-scale integration (MSI), which was more than 100, but less than 1 000 transistors on each IC. Next, with large-scale integration (LSI), more than 1 000 were put on each IC. By 1975, a great revolution occurred when more than 100 000 transistors were put on an IC, know as very-large scale integration (VLSI).

1.2 Electronic design

Electronic design applies to analogue or digital systems or to mixed analogue/ digital systems. The applied signal can range from low frequencies up to high-speed, high-frequency applications from DC up to frequencies exceeding 10 GHz. As with much of engineering design, there is no standard 'recipe' to follow. The design techniques used normally depends on the system requirements. Figure 1.1 outlines some of these, and include:

- electrical characteristics (currents, voltages and impedances);
- timing characteristics (rise/ fall times, clock speeds, propagation delays, and so on);
- temperature operating range (minimum and maximum operating temperatures, heat generated, and so on);
- power dissipation (drain on power supply, and so on);
- cross-talk (electromagnetic interference between devices and systems),
- functionality (programmable/fixed functionality, bus requirements, memory size, and so on);
- input/output characteristics (buffering, short-circuit protection, isolation, and so on);
- size of the package (number of pins, type of package, mechanical strength, and so on).

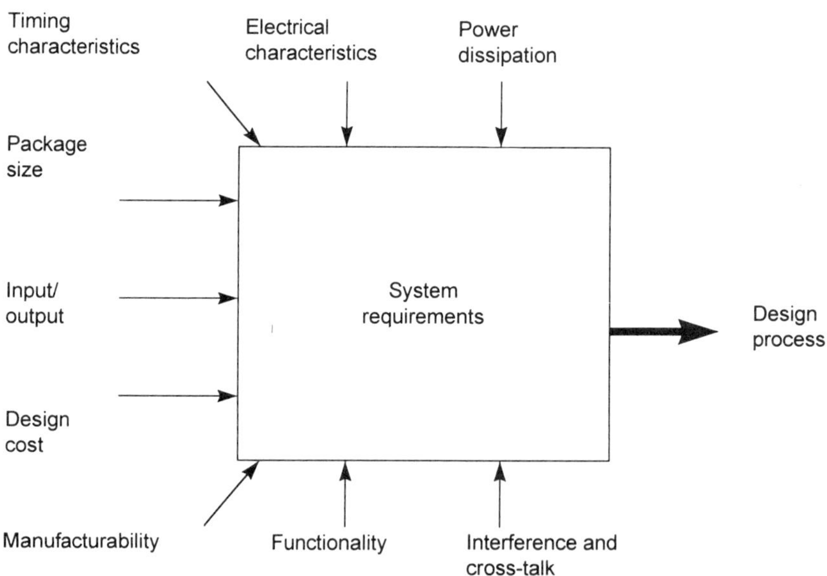

Figure 1.1 Typical system requirement.

Modern electronic design requires ever increasing functionality and lower package sizes. For example, a modern PC now contains an integrated microprocessor with an on-board memory cache, a 32-bit address bus, a 64-bit data bus, 64-bit registers, and so on. Functions, such as video control, power management systems and disk-drive controller integrate onto single ICs.

IC design has evolved from single transistors to SSI (small-scale integration), to MSI (medium-scale integration), to LSI (large-scale integration) and to VLSI (Very Large Scale Integration). An IC is normally classified either by the number of transistors it has, such as LSI, VLSI, and so on, or by the size of the transistor (covered in Chapter 4). Typical pitch sizes are 1, 1.5 and 2 μm (2 micros). Table 1.1 outlines the typical applications for the different classifications.

Table 1.1 Design classifications.

Type	*No. of transistors*	*Typical applications*
SSI	1-100	Logic gates, op-amps, linear applications.
MSI	100-1 000	Registers, filters, and so on.
LSI	1 000-100 000	8-bit microprocessors, up to 64 kbit ROMs and RAMs, Analogue-to-Digital converters, and so on
VLSI	100 000-500 000	16/32-bit microprocessors, up to 256 kbit ROMs/RAMs, signal processors.
ULSI†	>500 000	64-bit microprocessors, 8 Mbit RAMs, real-time and image processors.
GSI*	>10 000 000	64 Mbit RAMs, integrated multi-processors.

† ULSI represents ultra-large scale integration

* GSI represents gigantic scale integration

1.3 System development

System development normally involves the design of hardware, software and firmware, as illustrated in Figure 1.2. Hardware is the "parts that can be touched", such as electronic components, screws and nuts, the case, electrical wires, and so on. Software is the program that runs on processor hardware and changes its operation depending on the inputs. A program cannot exist without some form of hardware such as a microprocessor or a controller. Firmware are electronic devices that can be programmed using software. Typical firmware devices are EEPROMs (electrically erasable programmable read only memories), and interface ICs that are programmed using registers. In most applications, dedicated hardware responds faster than software running on processor hardware. But, systems running software tend to be easier to modify and require less development time.

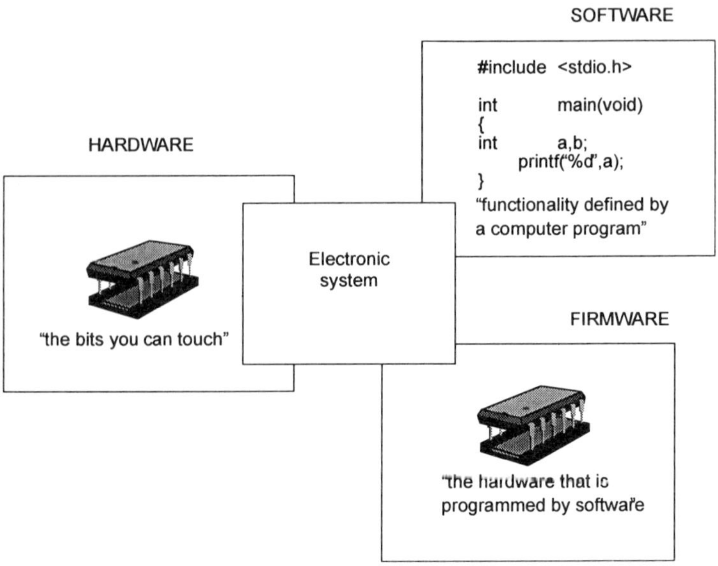

Figure 1.2 System hardware, software and firmware.

In modern electronic systems, hardware, software and firmware interconnect in the design process. Formal rules on the design of software have been developed over the years, these are now being applied to the development of hardware systems.

System development is normally a multi-staged process. It is iterative and a not sequential process and feedback from any part updates earlier stages. Figure 1.3 shows the main steps. At the first stage a need for the system is identified (the requirement analysis). If there is, a design team produces a design specification to define the functionality of the complete system.

After the production of the requirement specification, and when all parties involved find it acceptable, a full design takes place. This design solution must be implementable (such as manufacturable) and should comply exactly with the requirements specification. Next, an implementation of the design takes place using an appropriate technology.

The test, or validation, phase verifies that the implementation satisfies the original specification. System testing can only show the presence of errors and can never prove their absence. Normally it is necessary at this stage to produce test specifications and test data.

The last stage, the installation and maintenance phase, covers two quite distinct activities:

- to correct errors missed at an earlier stage but have been detected once the system is in operation;
- modifications due to additions or changes in the users requirement.

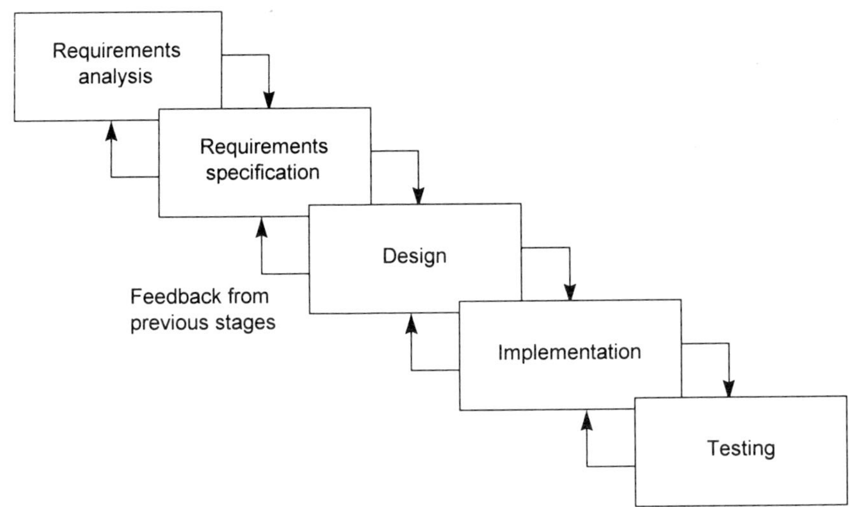

Figure 1.3 System development process.

1.4 Design abstraction

Electronic systems can be analyzed using a hierarchical structure or a layer approach. Figure 1.4 shows the main layers in this abstraction.

From this layered structure the design can either be approached, or viewed, from the top-down or from bottom-up. Top-down design is the most popular and initially involves defining the functionality of the system and how it interfaces with other systems. All inputs and outputs are identified and characterized. Next, at the functional layer, the system splits down into a number of discrete blocks. Each of these is either modeled as an algorithm or as a functional block with defined inputs and outputs. Next, at the circuit level, symbols joined by electrical connection represent each block. If specified by an algorithm then functional statements represent each block, or module. Below this layer, electrical symbols and connections (or algorithms) are represented pictorially by transistors and electrical routes. Finally, at the physical level the devices and connections are laid out in silicon. This involves the generation of several masks to define the IC layout.

Figure 1.5 shows the levels of abstraction and each layer can be used either in the design process, testing the circuit, fault finding, and so on.

A bottom-up approach is typically used when designing basic building blocks, such as programmable logic arrays (PLAs), memory arrays, multiplexers, decoders, registers, buses, and input/ output pads. From basic building blocks, a designer can build large systems such as arithmetic logic units (ALUs) or microprocessors. The most suitable design technique may involve a mixture of top-down and bottom-up approaches.

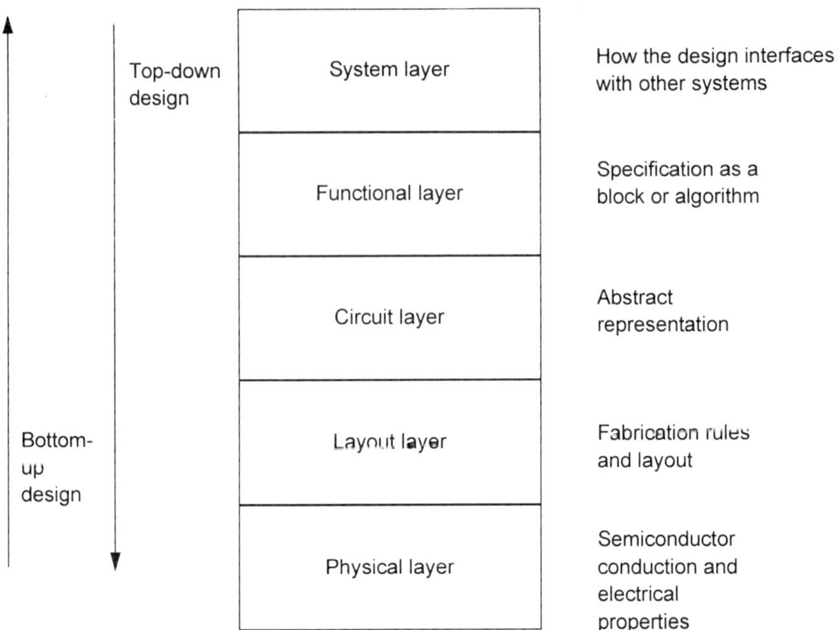

Figure 1.4 Design structure.

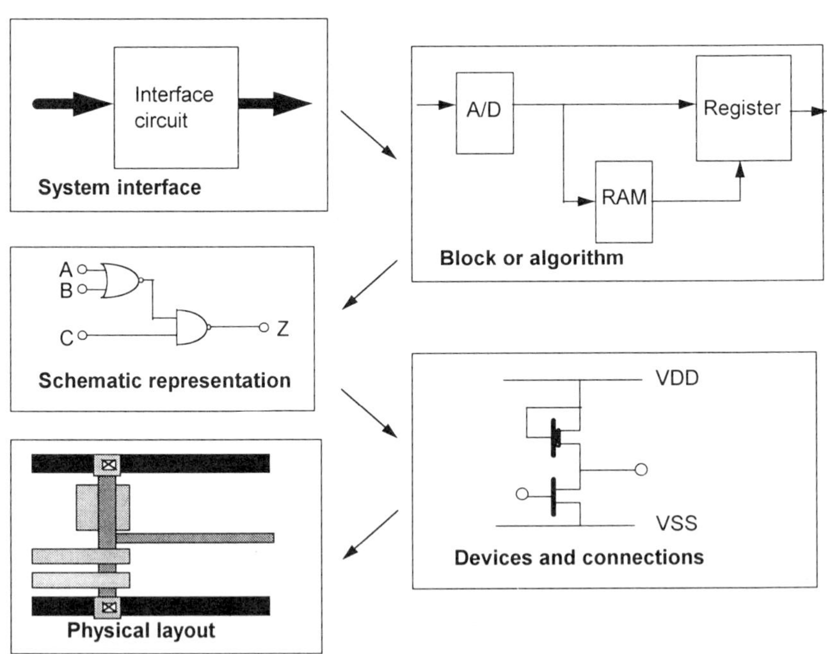

Figure 1.5 Levels of abstraction.

1.5 IC design process

For a top-down design, the designer produces sets of schematics or formal hardware descriptions. After this, a simulation of the design takes place to verify that the design complies with the original requirements. Most electronic computer-aided design (ECAD) packages produce sample test data which can be used to test the system when it is either placed on automated test equipment (ATE) or within a system. After validation the designer chooses between a full-custom or an automated layout design. For full-custom, the designer manually lays out, on silicon, the devices and routes within the design. In an automated layout the ECAD package uses a routing algorithm to place devices and then routes the connections between them. An automated design tends to be cheaper and more reliable, but may not meet all the requirements. Full-custom design is normally used when certain characteristics are to be met which cannot be satisfied by automated design tools. These are typically either a set of defined electrical characteristics or improved power dissipation. It can also be used to increase the density of components on the silicon. The final stage in the process is validation in which the IC is tested in isolation and then within the system. Figure 1.6 shows the main steps in this process.

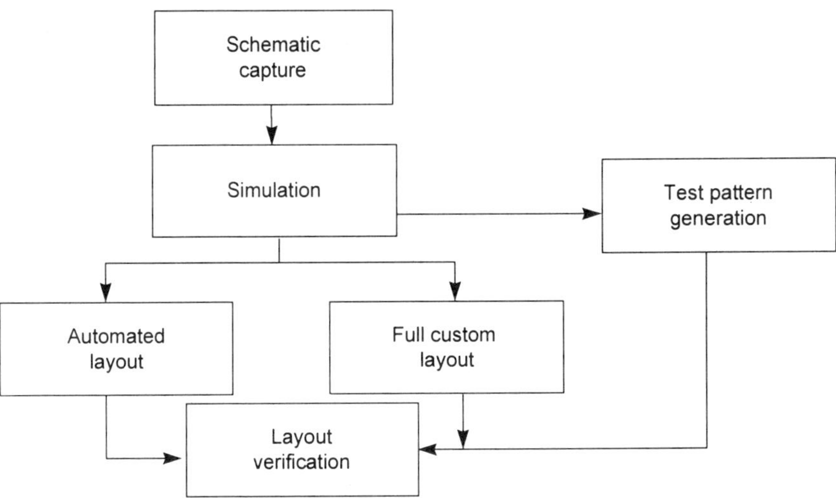

Figure 1.6 IC design process.

1.6 Logic technologies

There are two main types of semiconductor used in ICs: silicon (Si) and gallium arsenide (GaAs). The two main classes of transistor types are bi-polar and uni-polar. Bi-polar devices are normally used in high-speed and low noise applications. The main type of uni-polar device is MOS (metal-oxide semiconductor). These are typically used in low power applications and allow large densities. Figure 1.7 shows the main technologies used in digital design and Table 1.2 gives a basic description of these devices.

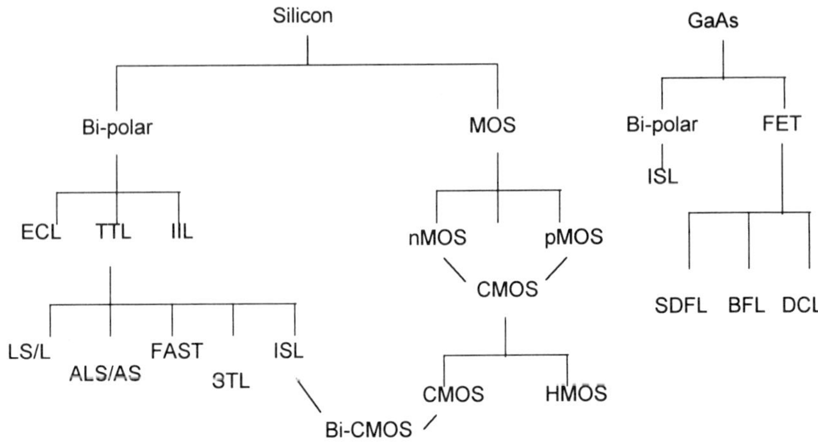

Figure 1.7 Digital IC technologies.

The dominant standard IC family for random logic used, for standard parts, is currently TTL (transistor-transistor logic). NMOS technology allows large densities of transistors and are typically used in microprocessors, signal processors and memories. CMOS is used in many low power applications, such as digital watches, hand-held games, and so on. Power dissipation in MOS devices normally depends on the clock frequency, that is, the higher the clock rate the higher the power dissipation. Many modern processors such as the Intel 80486 and Pentium microprocessors require large heat sinks because they operate at high clock rates.

Table 1.2 Logic device technologies.

Term	*Description*	*Comment*
ECL	Emitter coupled logic	Difficult to interface to, but are very fast bipolar devices.
TTL	Transistor-transistor logic	Standard logic devices series which is available in the 74-series.
IIL	Integrated injection logic (I^2L)	Difficult to interface to, but allows extremely high densities with low power requirements and good speed.

Table 1.2 (Cont.).

Term	*Description*	*Comment*
HMOS	High density MOS	Expensive, but allow high densities and low power usage.
NMOS	n-type MOS	Uses n-channel MOS devices which allows large densities and scaleable designs. It has relatively high power dissipation compared with CMOS and is used in memories, processors, and so on.
PMOS	p-type MOS	Uses p-channel MOS devices and is difficult to interface to as it requires 0 and –V power supply rails.
CMOS	Complementary MOS	Uses n- and p-channel transistors and has very low power dissipation but has relatively slow switching speeds. It is available for standard logic ICs with the 4000 series.
GaAs	Gallium arsenside	Very low switching speeds with clocking devices of several GHz. At present, it is difficult to fabricate with large transistor densities.
DCL	Direct coupled logic	A simple GaAs device, but is difficult to fabricate as it uses two types of transistors. It has a relatively low power dissipation compared with other GaAs logic types.
BFL	Buffered FET logic	A GaAs device which is easier to fabricate than DCL, but gives higher power dissipation.
SDFL	Schottky diode logic	A GaAs device which is easier to fabricate than DCL and lower power dissipation over BFL.

For the TTL series, the L represents low-powered TTL, S for schottky TTL, LS for low-powered schottky, FAST for advanced LS, ALS for advanced low-powered schottky TTL. Table 1.3 gives a comparison of the performance and characteristics of the logic families in terms of speed, densities and power dissipation.

Pre-fabricated designs are expensive in small volumes and have a fixed functionality but tend to be more efficient in their usage of the silicon and are less expensive in large volumes.

Devices which have their functionality pre-fabricated are known as mask-programmable and programmable devices as field-programmable.

Table 1.3 Logic device technologies.

Position	*Speed*	*Densities*	*Power dissipated*
1	GaAs (Fast)	I^2L (High)	I^2L (Low)
2	ECL	HMOS	CMOS
3	TTL	NMOS	NMOS
4	HMOS	CMOS	HMOS
5	NMOS	TTL	TTL
6	I^2L	GaAs	GaAs
7	CMOS (Slow)	ECL (Low)	ECL (High)

1.6.1 Full-custom design

Full-custom editing involves creating an IC by building each object manually. This technique usually allows very compact designs, but the efficiency of the design tends to depend on the skill of the designer. Modern ECAD tools and fabrication methods have reduced the need for full-custom design. It is typically only used to design very high-speed circuits or when ECAD packages using standard cell or gate array design cannot meet defined electrical parameters.

1.6.2 Programmable logic arrays (PLAs)

PLAs are either mask-programmable or field-programmable. They are made from arrays of AND and OR logic and are generally used to implement sets of Boolean equations. The OR logic implements the sum part of the equation and the AND logic the product part. Figure 1.8 illustrates the connections between the inputs and the AND array terms and the OR connections to the outputs.

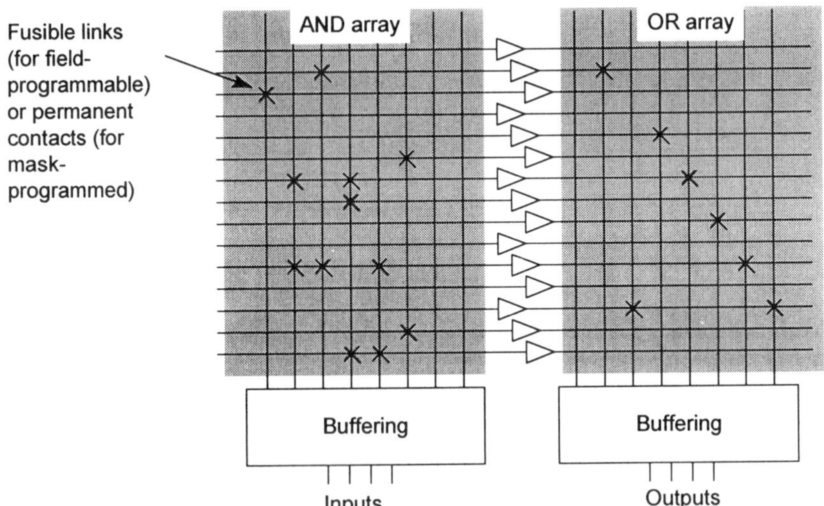

Figure 1.8 PLA structure.

If the PLA is mask-programmed, then the connections in the arrays are permanently set up when fabricated. When it is field-programmed then the functionality can either be made by fusing links or by making them electrically programmable. With fusible links, once the functionality is set up it is not possible to re-program the device. When electrically programmable, they can be re-program it as necessary.

Examples of fuse-link programmable devices are programmable read-only memories (PROMs), PLAs, and field programmable logic arrays (FPLAs). Examples of devices which can be re-programmed are electrically erasable devices electrically erasable programmable read-only memories (EEPROMs) and electrically programmed ROMS (EPROMs). EEPROMs are erased electrically, while EPROMs require ultra-violet light.

1.6.3 Programmable array logic (PALs)

PALs contain arrays of fixed logic functions, typically logic gates and flip-flops. The interconnections between the logic functions are permanent and are either fused or left as they are. In summary, they have:

- fixed array of logic cells;
- fixed connections between cells;
- fusible links break the connections between cells.

1.6.4 Gate arrays

Gate arrays are semi-customized ICs with a fixed structure of cells. Routing of one or more layers of metal customizes the design. They differ from PALs in that there are no fixed connections between transistor cells.

Gate array dies have collections of transistors placed in fixed positions together with suitable general-purpose peripheral input/output (I/O) structures and associated bonding pads. A characteristic feature of a gate array is the large number of potential internal contact points, and horizontal and vertical metal lines running in rows and columns across the device. Gate arrays are either fabricated with a fixed functionality or programmed by the user. Figure 1.9 shows an example gate array structure. A layer of insulation separates the two layers of metal. The solid and unfilled lines in the figure show the different metal layers and a square shows a via connection one layer to the next. For example Cell 1 connects to Cell 4 and Cell 15, and Cell 5 connects to Cell 13. Each of the cells has a fixed functionality thus some of the cells will not be required. In this case, Cells 2, 3, 6 and 7 are not used. This is of course inefficient in the usage of the silicon.

With a user programmable gate array, a gate array programmer sets up the required structure by fusing links on a metal layer(s). A typical programmable gate arrays is the Field Programmable Gate Logic Array (FPGLA).

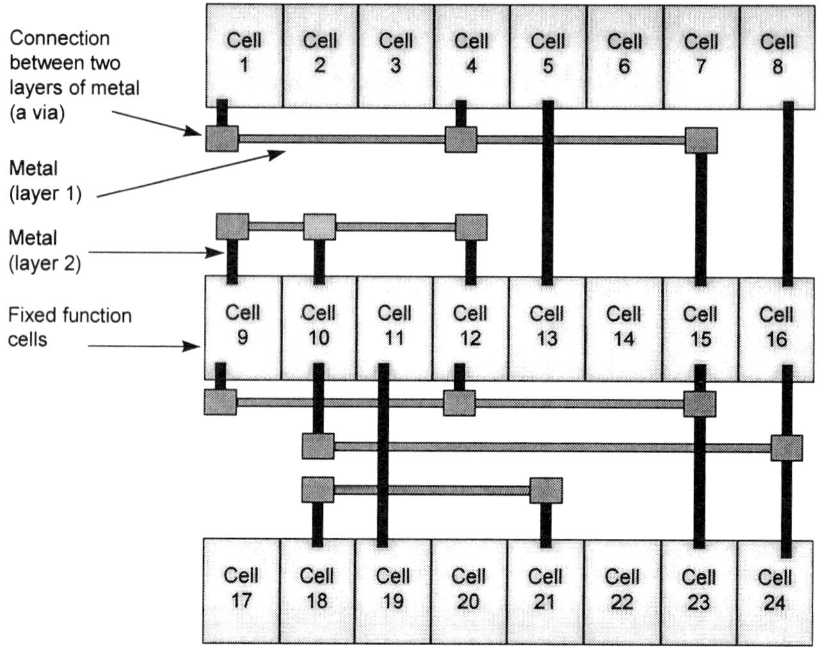

Figure 1.9 Gate array structure.

Field programmable arrays have the advantage over PLAs in that they contain features other than logic gates, such as input/output (I/O) devices and memory. Current technologies allow over 20 000 gates with 240 I/O lines. One advantage of field programmable devices is that hardware manufacturers can update systems without having to change any hardware. This could be used to provide new functions, bug fixes, hardware updates, and so on. A hardware manufacturer simply sends an update disk which downloads the required design onto the programmable gate array. They are typically the least expensive to design with and have very fast turn-around times, but are expensive in large volumes. In summary:

- transistors are arranged in fixed position in the array;
- transistors are unconnected before customization.

1.6.5 Semi-custom (standard cell)

Standard cell design involves storing standard transistor layout designs in a library and then interconnecting them to form customized circuits. The main difference between gate arrays and standard cells is that in standard cells the transistor layout and the routes between them are defined by the designer. Therefore, unlike a gate array, it is necessary to store all the transistor layouts and routes for each design. Typically there are more than 14 mask layouts, whereas gate array design normally only requires two layers to define the metal tracks.

Figure 1.10 shows an example design with 17 cells, which are interconnected by two layers of metal. For example Cell 1 connects Cells 3 and 11. Unlike gate array designs the cells are defined by the designer. These could be logic gates (AND, OR, NOR, NAND or EX-OR), flip-flops, memory cells, and so on.

There are a number of major arguments of advantages of standard cell over gate array designs, these include:

1. The ability to put an extra 20–30% (or more) functionality on a die, this can enable a system to be partitioned to minimize pin connections.
2. They are unlike gate arrays in that they do not have to accommodate for unlimited number of design elements. Thus elements which require large densities, such as microprocessors and memory elements, can be laid out with minimal areas.

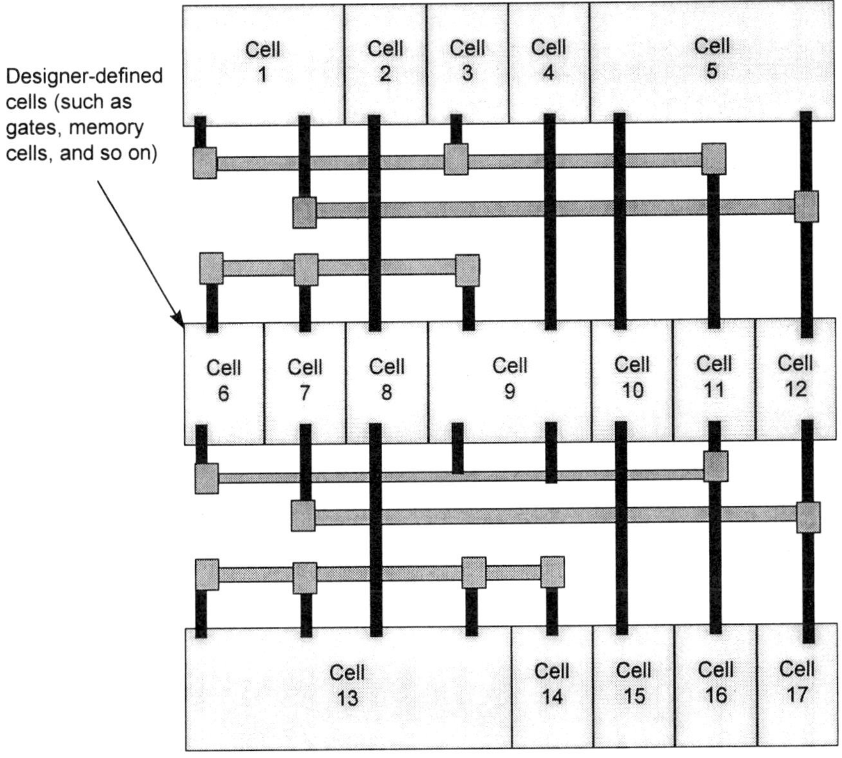

Figure 1.10 Semi-custom structure.

1.6.6 ASICs

Application Specific Integrated Circuits (ASICs) are designed for a specific user application. They typically contain mixed analogue/ digital devices with blocks such as ADCs and DACs, samplers, signal conditioners, memory, registers, processors, filters, and so on. By designing applications on a single IC, problems such as IC interconnection, multi-layered PCB design, cross-talk and interference can be reduced.

1.6.7 Comparison

Table 1.4 gives a comparison of the different design techniques. The type of technology used normally depends on the required volume of the IC and the number of gates/transistors in the design. Figure 1.11 shows that with large volumes and with a large number of gates the standard cell/ full-custom approach tends to be the most efficient solution. With programmable gate array there is a limit to the volume runs because they are relatively expensive to purchase and take time to program. When large volumes are required a mask-programmed gate array can be manufactured to the required functionality. This makes the device cheaper to manufacture in large volumes.

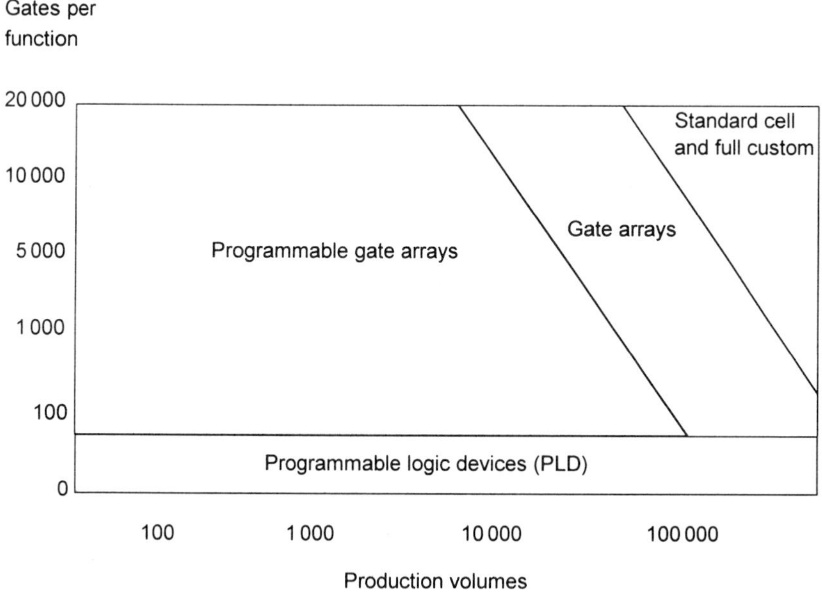

Figure 1.11 Comparison of differing design types.

Table 1.4 Comparison of the different design techniques.

	Advantages	*Disadvantages*
Full-custom	–Compact designs; –Improved electrical characteristics, such as speed optimizations or improved frequency response;	–Very time consuming in the design process; –Likely to be more error prone;
Semi-custom	–Designers/ vendors can build libraries of well-tested standard cells which can be shared between users; –Good for bottom-up design;	–Can be time consuming to built-up standard cell components; –Expensive in the short term but cheaper in long-term costs;
Gate array	–Fast implementation of designs and prototypes; –Easy updates; –Only two layers of metal require customization.	–Structure of IC is general purpose and thus can be wasteful of space and pin connections; –Relatively expensive in large volumes.

1.7 Jargon

A major obstacle in the communication of technological subjects is the usage of jargon. Table 1.5 outlines some of the terms used in microelectronic design.

Table 1.5 Typical terms used in IC design.

Term	*Description*
ASIC	Application Specific Integrated Circuit is an IC configured specifically for a particular application.
ATE	Automatic Test Equipment.
CAD	Computer-Aided Design is a general term applied to the use of computer packages in the design and analysis of products.
DRC	Design Rule Checker software checks the geometric definition of an IC. This includes the layout of the transistors and interconnected routes to ensure that they are within the required manufacturing tolerances.
Foundry	A foundry is the place where the silicon wafers are made.
Full-custom	A completely designed unit from the physical layout level.

Table 1.5 (Cont.).

Term	*Description*
HDL	Hardware Descriptor Language is a language used to described an electronic system in the form of an algorithm, such as VHDL.
Hybrid IC	A hybrid IC consists of a combination of two or more IC types or an IC with some discrete elements.
Integrated circuit	An integrated circuit (IC) is a combination of interconnected circuit elements inseparably associated on or within a continuous substrate.
Monolithic IC	A monolithic IC is an IC whose elements are formed in place upon or within a semiconductor substrate with at least one of the elements formed within the substrate.
PAL	Programmable Array Logic devices have arrays of logic.
PLA	Programmable Logic Array has an array of OR and AND logic.
Process	The 'recipe' used to manufacture a particular type of IC.
Semi-custom	A term applied to ASICs in which the IC layout information is customized to the requirements.
Standard cell	A pre-characterized functional unit used to build up larger functions in a cell-based design system.
Substrate	A substrate is the supporting material upon or within which an IC is fabricated or to which an IC is attached.
Wafer	A wafer is a circular slice of very pure silicon with repeated pattern of ICs. A typical production wafer may contain as few as 20 or 30 ICs or as many as several thousand, depending upon complexity and size of the fabricated circuit. Typically a wafer is circular and has a diameter of 4, 5, or 6 inches.

1.8 Exercises

1.1 If an IC has 50 000 transistors, determine its classification.

1.2 Contrast the different VLSI design techniques such as full-custom, standard cell and gate array.

1.3 Which is the fastest type of IC logic technology and which is the slowest.

1.4 Estimate, which type of IC logic technology offers the best performance in respect to speed and power dissipation.

1.5 Estimate, which type of IC logic technology offers the best performance in respect to speed and density.

1.6 Estimate, which type of IC logic technology offers the best performance in respect to speed, power dissipation and density.

1.7 Estimate, which type of IC logic technology would be preferable in digital watches.

1.8 How does the gate array design approach differ from semi-custom design.

1.9 Postscript: From the Bell Laboratories to Microsoft

In 1947 the invention of the transistor caused a storm. In fact, scientists at the Bell Laboratories kept its invention secret for over seven months so that they could fully understand its operation. On the 30 June 1948 the transistor was finally revealed to the world. Unfortunately, as with many other great inventions, it received little public attention and even less press coverage (*The New York Times* gave it 4½ inches on page 46).

Transistors had, initially, been made from germanium which is not a robust material and cannot withstand high temperatures. The first company to propose a method of using silicon transistors was a geological research company named Texas Instruments (which had diversified into transistors). Soon many companies were producing silicon transistors, and by 1955 the electronic value market had peaked, while the market for transistors was rocketing. The larger electronic valve manufacturers, such as Western Electric, CBS, Raytheon and Westinghouse failed to adapt to the changing market and quickly lost their market share to the new transistor manufacturing companies, such as Texas Instruments, Motorola, Hughes and RCA.

In 1959, IBM built the first commercial transistorized computer named the IBM 7090/7094 series. It was so successful that it dominated the computer market for many years. In 1965, they produced the famous IBM system 360 which was built with integrated circuits. Then in 1970 IBM introduced the 370 system, which included semiconductor memories. Unfortunately, these computers were extremely expensive to purchase and maintain.

Around the same time the electronics industry was producing cheap pocket calculators. The development of affordable computers happened when the Japanese company, Busicon commissioned a small, at the time, company named Intel to produce a set of eight to twelve ICs for a calculator. Instead of designing a complete set of ICs, Intel produced a set of ICs which could be programmed to perform different tasks. These were the first ever microprocessors. Soon Intel (short for *Int*egrated *El*ectronics) produced a general-purpose 4-bit microprocessor, named the 4004 and a more powerful 8-bit version, named the 8080. Other companies, such as Motorola, MOS Technologies and Zilog were soon also making microprocessors.

IBM's virtual monopoly on computer systems soon started to slip as many companies developed computers based around the newly available 8-bit microprocessors, namely MOS Technologies 6502 and Zilog's Z-80. IBM's main contenders were Apple and Commodore who introduced a new type of computer - the personal computer (PC). The leading systems where the Apple I and the Commodore PET. These spawned many others, including the Sinclair ZX80/ZX81, the BBC microcomputer, the Sinclair Spectrum, the Commodore Vic-20 and the classic Apple II (all of which where based on around the 6502 or Z-80).

IBM realized the potential of the microprocessor and used Intel's 16-bit 8086 microprocessor in their version of the PC. It was named the IBM PC and has since become the parent of all the PCs ever produced. IBM's main aim was to make a computer which could run business applications, such as word processors, spreadsheets and databases. To increase the production of this software they made information on the hardware freely available. This resulted in many software packages being developed and helped clone manufacturers to copy the original design. So the term 'IBM-compatible' was born and it quickly became an industry standard by sheer market dominance.

On previous computers IBM had written most of their programs for their systems. For the PC they had a strict time limit, so they went to a small computer company called Microsoft to develop the operating system program. This program was named the Disk Operating System (DOS) because of its original purpose of controlling the disk drives. It accepted commands from the keyboard and displayed them to the monitor. The language of DOS consisted of a set of commands which were entered directly by the user and interpreted to perform file management tasks, program execution and system configuration. The main functions of DOS were to run programs, copy and remove files, create directories, move within a directory structure and to list files.

Microsoft has since gone on develop industry-standard software such as Microsoft Windows Version 3, Microsoft Office and Microsoft Windows 95. Intel has also benefited greatly from the development of the PC and has developed a large market share for their industry-standard microprocessors, such as the 80286, 80386, 80486 and Pentium processors.

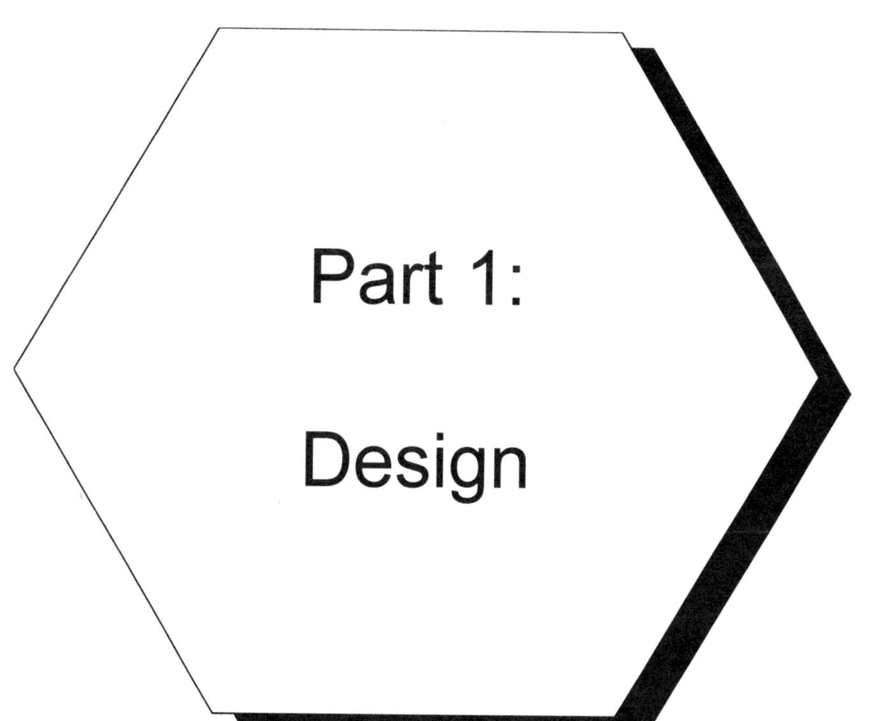

Part 1:

Design

Bi-polar Design

2

2.1 Introduction

The four main device technologies used in electronic logic design are TTL, CMOS, NMOS and Bi-CMOS. This and the next chapter discusses their operation and compares their characteristics. Transistor-transistor logic (TTL) uses bi-polar transistors and as a consequence they are generally resistant to damage. Unfortunately, when used as a switch, a bi-polar transistors' signal path resistance varies with the applied signal voltage, this leads to signal distortion. It also cannot pass signals in both directions. MOS devices overcome these limitations.

In analyzing bi-polar circuits it assumed that a fully conducting diode junction has a voltage drop of approximately 0.7 V. For an npn transistor, if a significant bias current flows from the base to the emitter then the base-emitter voltage can be assumed to be 0.7 V. It is also assumed that the collector-emitter voltage of the saturated npn transistor is approximately 0.2 V. These approximations are illustrated in Figure 2.1

2.2 TTL design

Transistor-transistor logic (TTL or T^2L) is one of the most widely used logic technologies. Figure 2.2 shows a TTL inverter (or NOT) circuit. There are three main stages to this circuit. The first uses transistor Q_1 as an input coupling stage.

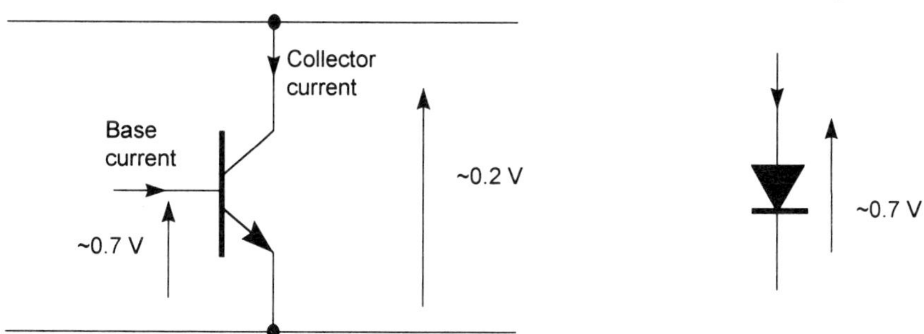

Figure 2.1 Saturated bi-polar npn transistor.

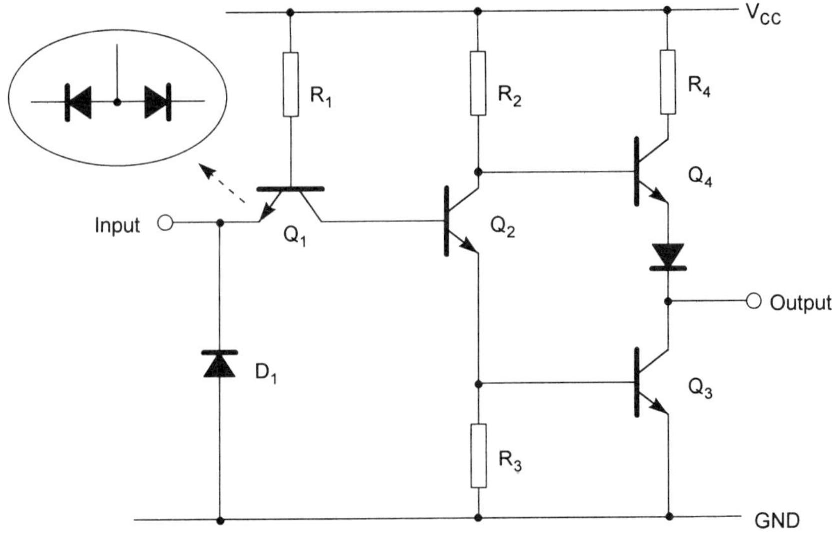

Figure 2.2 TTL inverter circuit.

From this stage, the output feeds into a phase-splitter circuit using Q_2. Finally, a totem-pole arrangement drives the output using Q_3 and Q_4.

The input clamp diode D_1 protects against excessive negative input voltages. When the input voltage is below –0.6 V the clamp diode conducts and limits the input voltage to around –0.6 V. It thus prevents negative spikes on the input voltage damaging Q_1.

A HIGH input causes the base-emitter junction of Q_1 to be reverse-biased, and the base-collector junction to be forward-biased. Current then flows through R_1 into the base-collector junction of Q_1, then into base of Q_2, thus driving Q_2 into saturation. As a result, Q_3 is ON due to the current supplied by Q_2. Its collector voltage, which is the output, will be near ground potential (~0.2 V). The collector of Q_2 is at a sufficiently low voltage level to keep Q_4 OFF. Thus a HIGH input causes a LOW output.

For a LOW input the base-emitter junction of Q_1 is forward-biased and the base-collector junction reversed-biased. The current then flows through R_1 into the base-emitter junction of Q_1 and then out into the Input connections. No current then flows into the base of Q_2, so it is OFF. The collector of Q_2 will thus be HIGH, turning Q_4 ON. This transistor provides a low impedance path from V_{CC} to the output. The emitter of Q_2 is at ground potential keeping Q_3 OFF. Thus a LOW input causes a HIGH output. It is important to note that a LOW input must thus provide a path for current from the load gate to ground.

Diode D_2 ensures that Q_4 will be OFF when Q_2 is ON (HIGH input). The collector voltage of Q_2 is equal to the V_{BE} of Q_3 plus the V_{CE} of Q_2. Diode D_2 provides an additional V_{BE} (0.7 V) equivalent drop in series with the base-emitter junction of Q_4 to ensure that it is OFF when Q_2 is ON.

Figures 2.3 and 2.4 shows the approximate voltages for the TLL inverters for each of the input states. In Figure 2.3 the base of Q_1 is 2.1 V because of the three conducting diode voltage drops, that is $V_{BE}(Q_3)$, $V_{BE}(Q_2)$ and $V_{BE}(Q_1)$. This voltage level causes Q_2 and Q_3 to be ON.

In Figure 2.4 the input voltage is assumed to be 0 V thus the base of Q_1 is approximately 0.7 V which is not a high enough voltage to turn Q_2 and Q_3 ON.

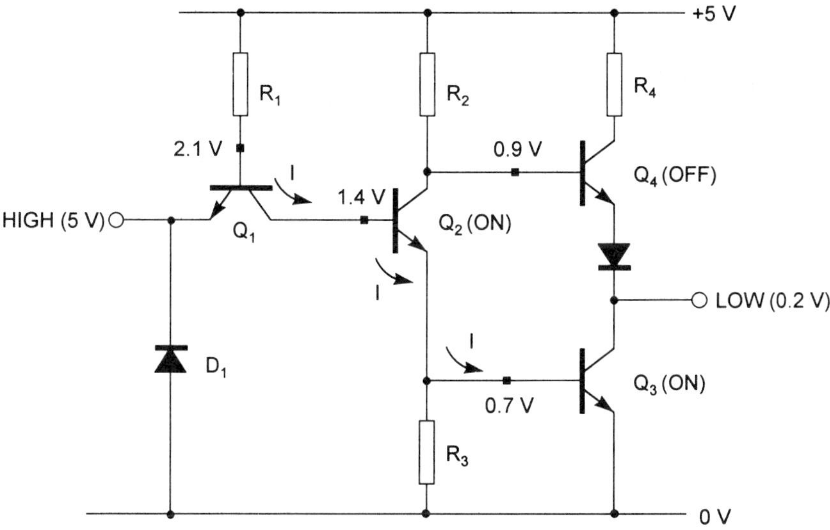

Figure 2.3 TTL inverter operation (HIGH input).

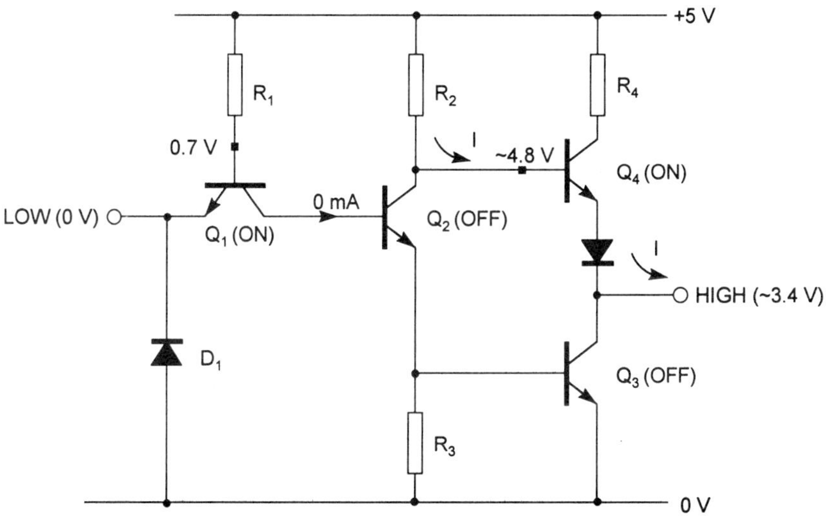

Figure 2.4 TTL inverter operation (LOW input).

Figure 2.5 shows a two-input TTL NAND gate. It uses multiple-emitter transistors to couple the input connections. The insert shows the equivalent diode arrangement for a dual-emitter transistor. When the inputs A and B are HIGH then the base-emitter junction of Q_1 is reversed biased and the base-collector junction conducts. This causes Q_2 to be ON and provides current for Q_3. The output will thus be LOW.

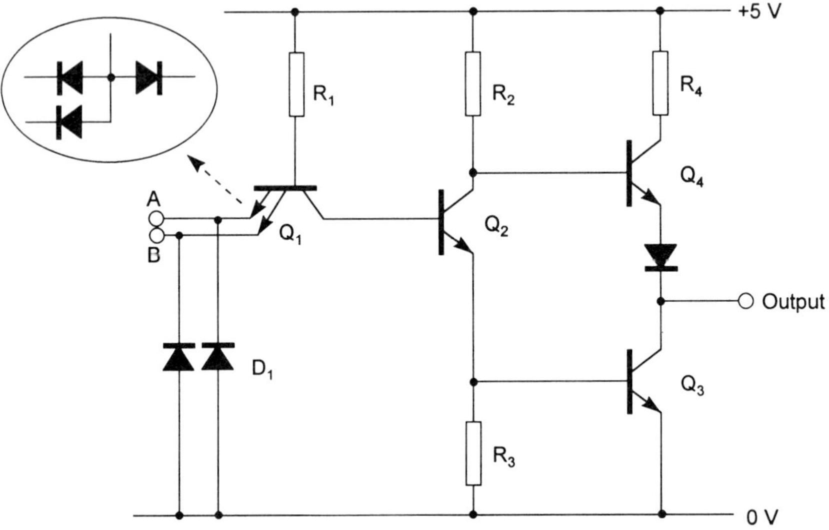

Figure 2.5 TTL NAND gate circuit.

If any of the inputs is LOW then one of the base-emitter junctions of Q_1 will be forward biased. This causes Q_2 and Q_3 to be OFF and Q_4 ON. The output will thus be HIGH.

It is important to realize that the output of a TTL gate output sinks current when the output is LOW and sources a small current when the output is HIGH. Figure 2.6 shows the resistor values used in a standard 54/74 series NAND gate.

The short circuit current for the gate in Figure 2.6 can be determined from the voltage drop across the 130 Ω resistor, which will be approximately $V_{CC} - V_{CE}(Q_4) - V_D\,(D_2) = 5 - 0.2 - 0.7 = 4.1$ V. The short-circuit current will thus be this voltage divided by 130 Ω, or 31.5 mA.

Figure 2.7 shows the low-powered 74L series of TTL gates. To reduce current flowing in the circuit the resistor values are increased. A reduction of current leads to less power dissipation. A typical 54/74 gate power dissipation is around 10 mW whereas an equivalent 54L/74L dissipates around 1 mW. Unfortunately, this increased resistance increases pulse rise times and leads to a loss of speed. For example, a typical standard 54/74 gate has a propagation delay of 10 ns compared with 33 ns for a 54L/74L gate.

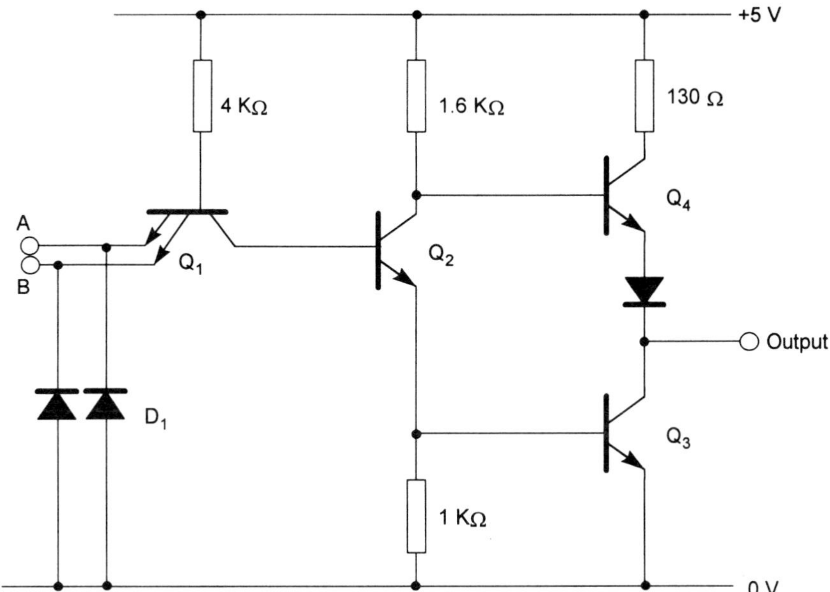

Figure 2.6 Standard TTL NAND gate.

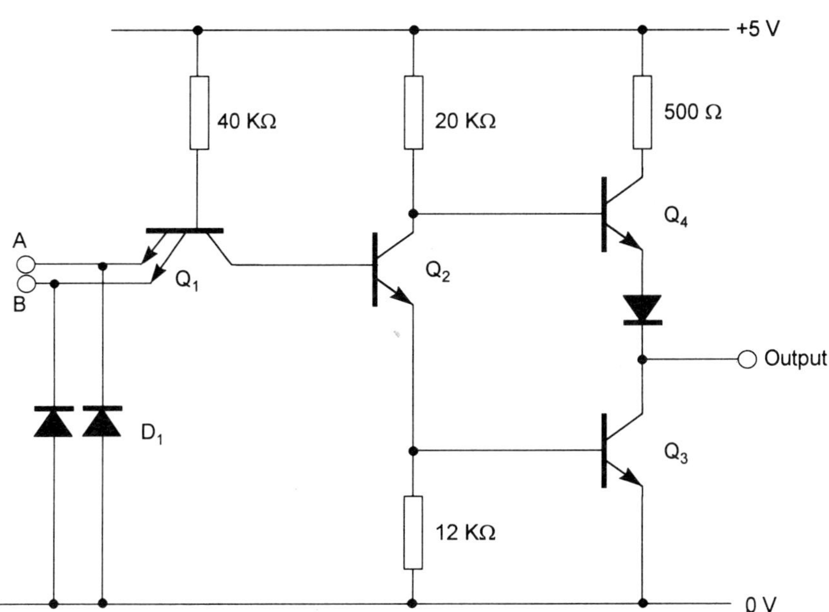

Figure 2.7 TTL 54L/74L series NAND gate.

Figure 2.8 shows a high-speed TTL (54H/74H) circuit which provides lower propagation delays than the standard gates, but result in a high power dissipation.

The 54S/74S series provides even faster switching times than the equivalent 54H/74H but requires much less power. These circuits incorporate Schottky diodes which prevent the transistors saturating. This decreases the time for a transistor to turn ON or OFF. Typical propagation delays for a 54S/74S gate is 3 ns and the power dissipation is around 19 mW. Table 2.1 outlines typical TTL performance characteristics.

Table 2.1 Typical characteristics of different TTL technologies.

Series	*Propagation delay*	*Power dissipation*	*Max. clock frequency*
54/74	10 ns	10 mW	35 MHz
54L/74L	33 ns	1 mW	3 MHz
54H/74H	6 ns	22 mW	50 MHz
54S/74S	3 ns	19 mW	125 MHz
54LS/74LS	9.5 ns	2 mW	45 MHz

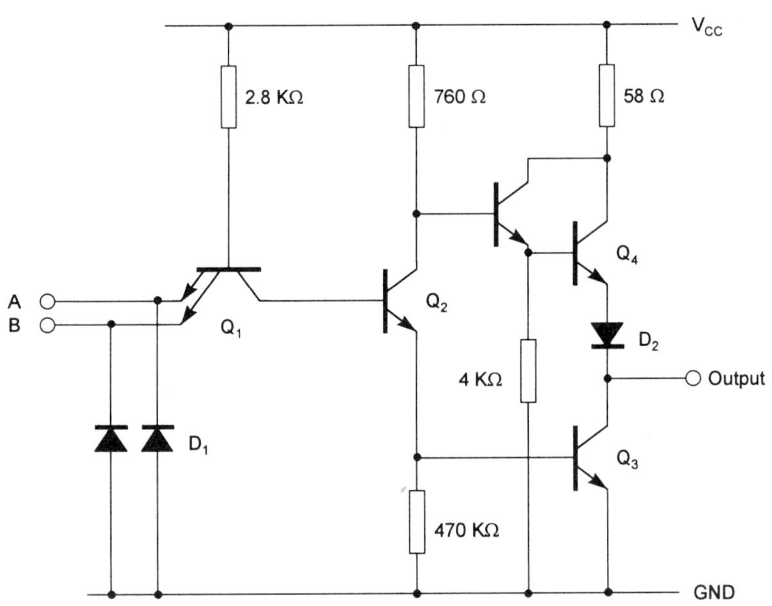

Figure 2.8 TTL 54H/74H series NAND gate.

2.2.1 TTL NOR gate

Figure 2.9 shows a 2-input NOR gate. When both inputs are a LOW then the Q_1 and Q_2 base-emitter junctions are forward-biased thus there is no bias current for Q_3 and Q_4. The output transistor Q_5 will thus be ON and the output will be

HIGH. If the A input is LOW then the base-emitter junction of Q_1 is reversed biased and the base-collector diode provides bias current to Q_4. If the B input is LOW, then Q_2 provides bias current to Q_3. Thus a HIGH on any of the inputs biases either Q_3 and/or Q_4. These then take bias current away from Q_5 and supply it to Q_6. This causes a LOW output.

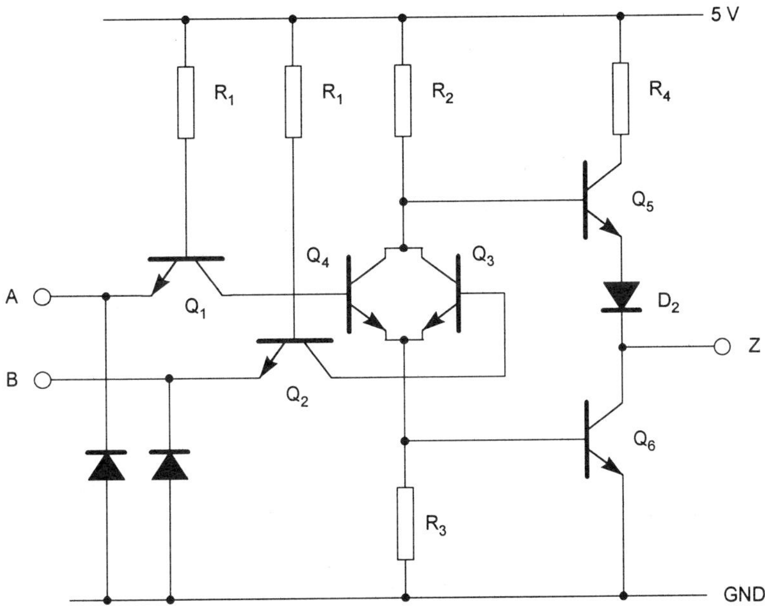

Figure 2.9 2-input TTL NOR gate.

2.2.2 Fan-out

TTL gates rely on the flow of current from/into the input of one gate into/from the output of another. Unfortunately, there is a limit to the amount of current a transistor can source or sink and still reliably provide the required voltage levels. Gate fan-out is the maximum number of inputs, from devices of the same IC family, that the gate can drive whilst maintaining its output levels within specified limits. Thus the fan-out specifies the maximum loading that a given gate can handle.

The output stage from a TTL gate requires to sink current with a LOW output state (that is, current flows into the driving gate). When the driving gate output is HIGH then current flows out of the gate, thus the output stage sources current.

When the output from the driving gate is HIGH then the input to the load gate is like a reversed biased diode and there will be practically no current required from the driving gate. In a typical TTL gate a 40 μA reverse leakage current flows from the driving gate into the load gate input, as shown in Figure 2.10.

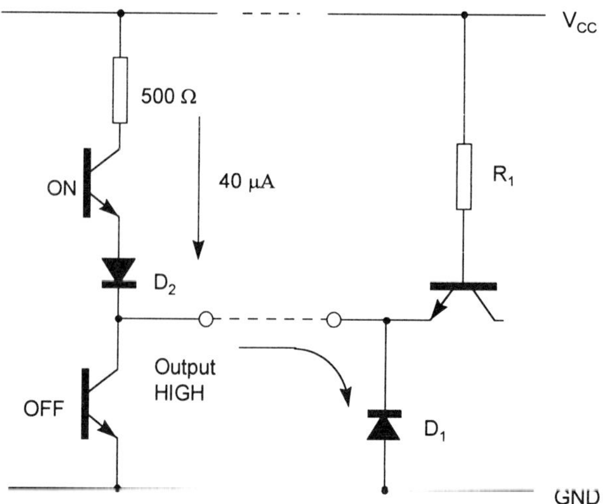

Figure 2.10 Driving a TTL input from output HIGH state.

When the output from the driving gate is LOW then the input to the loading gate is like a forward biased diode. A current then flows out of the loading gate into the driving gate output, as shown in Figure 2.11. The driving gate then sinks current from the loading gate. For a standard TTL gate this current is about 1.6 mA and can be approximated by determining the voltage across R_1, which will be $V_{CC} - V_{BE}(ON) - V_{CE}(ON)$. The current will thus be:

$$I_{OL} = \frac{5 - 0.7 - 0.2}{R_1} \quad A$$

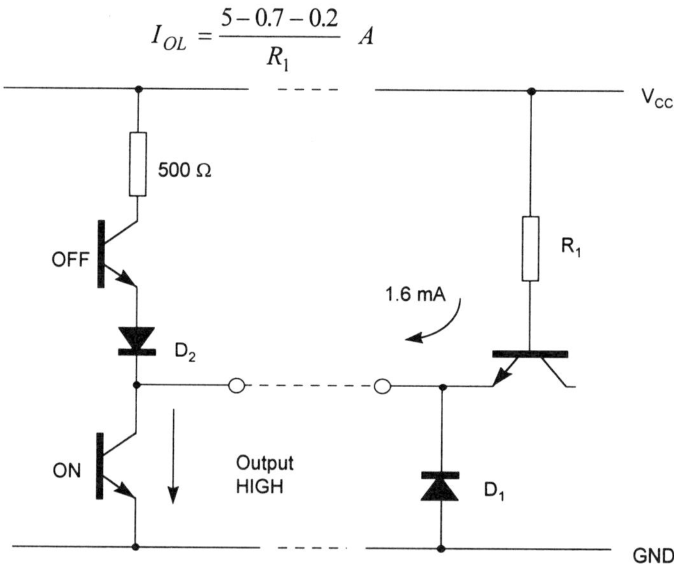

Figure 2.11 Driving a TTL input from output LOW state.

One TTL input is termed a unit load (UL) and, for an output LOW, represents a sink current of 1.6 mA. The maximum number of unit loads that a typical TTL gate can handle is called its fan-out.

A standard TTL gate has a fan-out of 10 UL (except for buffer gates, which have a fan-out of 30 UL). Thus a standard TTL gate can connect to up to 10 other standard gate inputs. This represents a current of around 16 mA. Table 2.2 gives typical electrical characteristics of a TTL gate.

Table 2.2 Typical TTL characteristics.

	Min	*Typical*	*Max.*	*Description*	*Test conditions*
V_{IH}	2.0 V			Input HIGH voltage	Guaranteed HIGH input
V_{IL}			0.8 V	Input LOW voltage	Guaranteed HIGH input
V_{OH}	2.5 V	3.3 V		Output HIGH voltage	I_{OH} =0.4 mA, V_{IN}=0.8 V
V_{OL}		0.22 V	0.4 V	Output LOW voltage	I_{OH} =0.4 mA, V_{IN} =0.8 V
I_{OH}			–0.4 mA	Maximum output HIGH current	
I_{OL}			16 mA	Maximum output LOW current	
I_{IH}			40 μA	Maximum input HIGH current	V_{IN} =2.4 V
I_{IL}			–1.6 mA	Maximum input LOW current	V_{IN} =0.4 V

The standard TTL unit load is thus:

1 TTL Unit Load (UL) = 40 μA in the HIGH state;
1 TTL Unit Load (UL) = 1.6 mA in the LOW state.

EXAMPLE (output load):

1. The output of a 7400 gate sinks a maximum of 16 mA in a LOW state and sources a maximum of 800 μA in the HIGH state. For a LOW output the maximum driving factor will thus be:

$$\frac{16\text{ mA}}{1.6\text{ mA}} = 10\text{ UL}$$

and for a HIGH output the drive factor is:

$$\frac{800\ \mu\text{A}}{40\ \mu\text{A}} = 20\ \text{UL}$$

The worst-case loading thus occurs when the output is a LOW as the LOW state can only interface to 10 unit loads while in the high state the output can drive 20 unit loads. Thus the unit load for the gate will be the lowest value of the two, that is 10 UL.

2. The output of the 74LS00 gate sinks a maximum of 8.0 mA in the LOW state and sources a maximum of 400 μA in the HIGH state. For a LOW output the maximum driving factor will thus be:

$$\frac{8.0\ \text{mA}}{1.6\ \text{mA}} = 5\ \text{UL}$$

and for an output HIGH the drive factor is:

$$\frac{400\ \mu\text{A}}{40\ \mu\text{A}} = 10\ \text{UL}$$

Table 2.3 gives sample unit loads for different families of TTL gates.

Table 2.3 Typical unit loads.

Device	*HIGH drive*	*LOW drive*	*Device*	*HIGH drive*	*LOW drive*
74LS00	10 UL	5 UL	74S00	25 UL	12.5 UL
7400	20 UL	10 UL	74FAST	25 UL	12.5 UL
74H00	25 UL	12.5 UL	74ALS	10 UL	5 UL

Unconnected inputs of a TTL gate act as a HIGH as there is no bias current for the coupling transistor, acting like a HIGH input level. However, with added noise or the coupling of external electrical paths, a small current could cause the input to act as a LOW. It is thus best not to leave unused TTL inputs unconnected (or open) and do one of the following with unused inputs:

1. Connect unused inputs to used inputs, if the maximum fan-out of the driving gate is not exceeded.
2. Connect unused inputs of AND and NAND gates to the DC supply voltage through a 1 KΩ resistor and connect unused inputs of OR or NOR gate to ground.

2.3 I^2L design

Integrated injection logic (I^2L) is a bi-polar technology that gives extremely high component densities (up to 10 times that of TTL) and reasonably fast switching speeds. It is also simpler to fabricate than CMOS, NMOS and TLL and has low power requirements.

Figure 2.12 shows a basic I^2L gate. Transistor Q_1 acts as a constant current source because there is a constant base-emitter voltage (assuming there is no variation in temperature). The multiple-collector transistor operates as an inverter. The great advantage of the I^2L gate is that it only requires the space of a single TTL multiple-emitter transistor. This is because the base of Q_1 and the emitter of Q_2 connect together and they both connect to GND. The collector of Q_1 and base of Q_2 are also common.

Transistor Q_1 is the current-injector transistor and supplies a constant current to base of Q_2. The current is constant because the base-emitter voltage of Q_1 is fixed as the emitter connects to a fixed voltage (~0.6 V). A LOW on the base of Q_2 takes the injection current away from the base of Q_2 and through the low impedance path provided by the driving gate. This turns Q_2 OFF. For a HIGH input the driving gate will be high impedance and thus the injector current flows into the base of Q_2, which turns it ON.

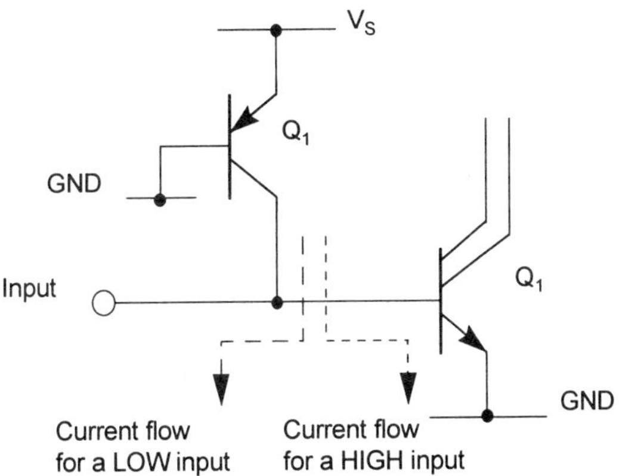

Figure 2.12 I^2L gate.

2.4 ECL design

ECL stands for emitter-coupled logic and uses bi-polar devices. A typical OR/NOR gate is given in Figure 2.13. It consists of three main stages: the differential amplifier input stage, a bias circuit to produce a threshold voltage and an emitter-follower output stage. The output stage provides automatic logic complement (that is OR and NOR).

One problem with TTL circuits is that they require a relatively large change in input voltage before the circuit state changes. In standard TTL the maximum voltage for a LOW is 0.8 V and the minimum voltage for a HIGH is 2.5 V (refer to Table 2.2). Thus when changing state the voltage must change from 0.8 V to 2.5 V. For increased switching speed a better solution is to have a small change-over voltage range and for the voltage representing the two states to be set at a voltage just outside this range. ECL does this by amplifying the difference between the input voltage level and a fixed threshold voltage. Another disadvantage with TTL circuits is that transistors operate in saturation which means that they take a longer time to switch between states. ECL circuits have the advantage in that none of the transistors saturate. Unfortunately, because the transistors always conduct, no matter what state they arc in, there is an increase in power dissipation.

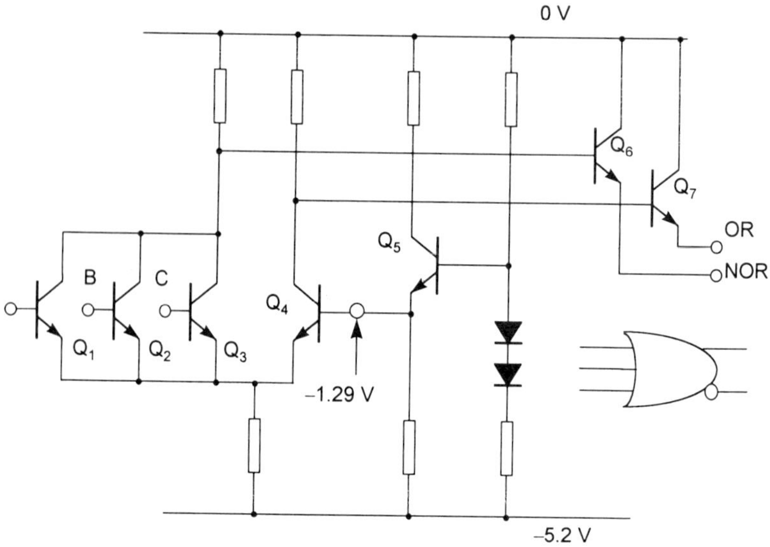

Figure 2.13 An ECL OR/NOR gate circuit.

A typical ECL gate connects the V_{CC} pin to ground, and the V_{EE} to −5.2 V. Figure 2.14 shows a typical switching characteristic. The output varies from a LOW level of −1.75 V to a HIGH of −0.9 V. In positive logic, a 1 is the HIGH level (less negative) and a 0 is the LOW level (more negative).

A LOW on all the inputs (typically −1.75 V) causes Q_1 to Q_3 to be OFF because their base-emitter junctions are reverse-biased. Transistor Q_4 conducts but will not saturate. The bias circuit holds the base of Q_4 at −1.29 V, and therefore its emitter is approximately 0.7 V below the base, at −1.99 V. For a low input the voltage differential from base to emitter of the input transistors Q_1 to Q_3 is −1.99–(−1.75)= −0.24 V, which is less than the required forward-bias voltage and these transistors will therefore be OFF, as shown in Figure 2.15.

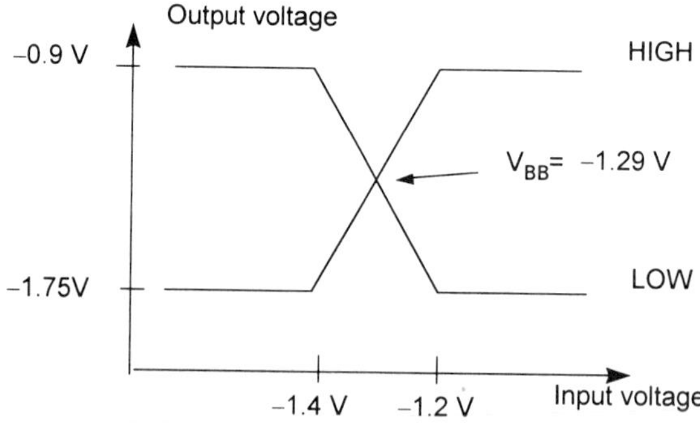

Figure 2.14 Gate transfer curves.

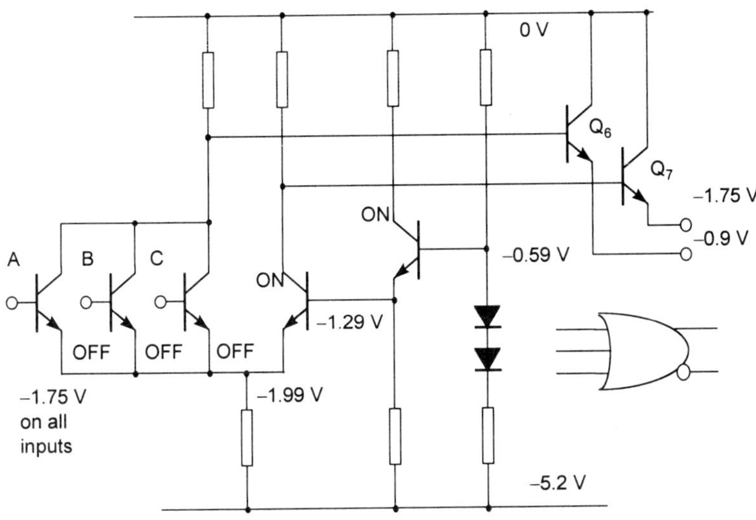

Figure 2.15 Operation when inputs are LOW.

When any one (or all) of the inputs is HIGH (–0.9 V) then the input transistor(s) conducts. The voltage at the emitters of Q_1 to Q_4 increases from –1.99 V to –1.6 V (one base-emitter drops below the –0.7 V). Since the base of Q_4 is held at a constant –1.29 V by the bias circuit then Q_4 turn OFF. The resulting collector voltages feed through the emitter followers to the output. Because of the differential action of Q_1 to Q_3 and Q_4, Q_4 is OFF when one or more of the Q_1 to Q_3 transistors conduct. This provides simultaneous complementary outputs as shown in Figure 2.16.

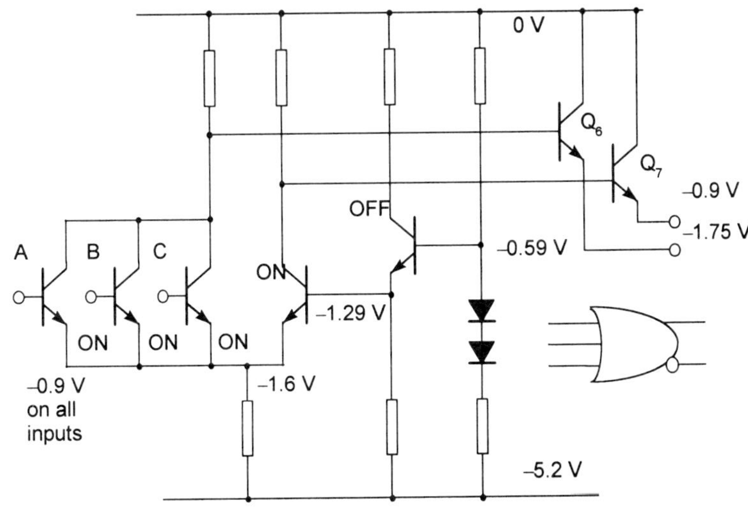

Figure 2.16 Operation when inputs are HIGH.

2.4.1 Design of high-speed logic

High speed operation involves only four considerations that differ significantly from operation at low and medium speeds.

1. The time delay through interconnected wiring are highly important.
2. The possibility of distorted waveforms dues to reflections on mismatch signal lines.
3. The possibility of crosstalk between adjacent signal lines.
4. Increased amount of electrical noise generation and pick-up.

The interconnection-wiring delays can be reduced by reducing the length of interconnected lines. Waveform distortion due to line reflections can be reduced by matching all line and input/ output impedances to the same impedance, such as 50 Ω.

2.4.2 ECL design example

Figure 2.17 shows a user designed ECL gate. Diodes D_1 and D_2 simulate the voltage drops across the base-emitter junctions on Q_5 and Q_4. When the input level is LOW then the voltage across the 150 KΩ resistor will thus be the same as the voltage drop across the 400 Ω emitter resistor. This provides a constant current source for the differential stage. The threshold voltage will thus be V_A+0.7 V. The voltages and currents for a HIGH input can be determined by:

$$V_{150K} = \frac{5.2 - 0.7 - 0.7}{2} = 1.9 \text{ V}$$

$$V_C = -5.2 + 1.9 + 0.7 + 0.7 = -1.9 \text{ V}$$

$$V_B = -2.6 \text{ V}$$

$$V_A = -3.3 \text{ V}$$

$$I_{400R} = \frac{V_{400R}}{400} = \frac{5.2 - 3.3}{400} = 4.75 \text{ mA}$$

$$I_{15K} = \frac{V_{15K}}{15K} = \frac{5.2 - 2.6}{15K} = 0.17 \text{ mA}$$

$$I_{150K} = \frac{V_{150K}}{150K} = \frac{0 - 1.9}{150K} = 12.7\ \mu\text{A}$$

$$V_D = 4.7\text{mA} \times 400\text{R} = -1.9 \text{ V}$$

$$V_O = -1.9 - 0.7 = -2.6 \text{ V}$$

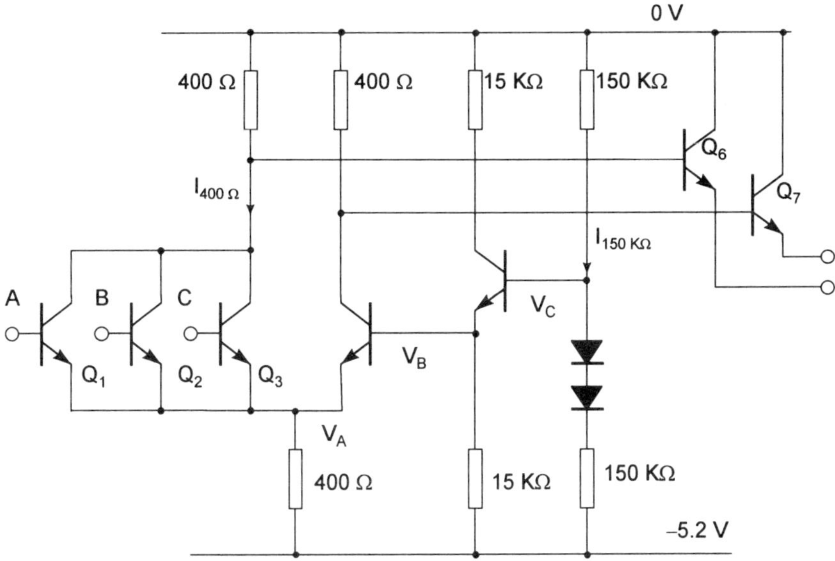

Figure 2.17 User-designed ECL gate.

2.4.3 ECL advantages and disadvantages

The advantages of ECL are:

1. Very high switching speed because the transistors do not saturate.
2. They provide automatically complementary outputs.
3. Smooth switching of transistors reduces switching spikes on the power supply.
4. Outputs from gates connected together to give implied-OR function.
5. There is little degradation of parameters with variations in temperature.
6. Number of available logic functions high.
7. High fan-out because of high input and low output impedance.
8. Provides data transmission over long distances by means of balanced twisted-pair 50 Ω lines. In general, balanced lines are less prone to noise than single-ended.

The disadvantages are:

1. Small voltage difference (800 mV) exists between the two logic levels and noise margins are only +/–200 mV.
2. Relatively high power dissipation.
3. Difficult to interface with other logic families as this requires some form of level shifter.
4. Capacitive loading reduces the speed of operation, and thus the fan-out.

2.5 Exercises

2.1 Explain why an open-circuited TTL input acts as a HIGH input.

2.2 If two unused inputs of a TTL gate connect to an input being driven by another TTL gate, how many other inputs can be driven by this gate. Assume the fan-out is 10 unit loads.

2.3 With reference to Table 2.4, determine how many 74LS gates a standard gate can drive.

2.4 With reference to Table 2.4, determine how much current a standard 74-series TTL gate sinks if it is driving seven 74S loads.

2.5 If a standard TTL gate drives six unit loads, determine the amount of current it sinks in the LOW output state.

2.6 If a standard TTL gate drives 10 unit loads, determine the amount of current it sources in the HIGH output state.

2.7 Estimate the currents and voltages in the TTL circuits in Figure 2.6 and also for the circuit in Figure 2.7.

2.8 The circuit given in Figure 2.18 is made up of 7400 and 74LS00 gates. Determine the maximum currents for I_1, I_2, and I_3 for a LOW and a HIGH input, highlighting any problems that may occur (refer to Table 2.4).

2.9 Determine the logic function for the TTL gate in Figure 2.19.

2.10 Design a TTL gate, with the minimum of transistors, for the following:

(a) $Z = \overline{(A.B) + (C.D) + (E.F)}$ (b) $Z = \overline{A.(E + F) + B.C}$

2.11 A user-defined ECL gate is shown in Figure 2.20. Explain how the diodes D_1 and D_2 set up a constant current through the lower 300 Ω resistor in the differential input stage. Also determine the voltages V_A, V_B, V_C, V_D and V_O for LOW and HIGH inputs.

Table 2.4 Typical unit loads.

	I_{IL}	I_{IH}	I_{OL} *(max)*	I_{OH}*(max)*
Standard	−1.6 mA	40 μA	16 mA	−0.4 mA
LS-series	−0.4 mA	20 μA	8 mA	−0.4 mA
S-series	−2 mA	50 μA	20 mA	−1 mA

2.12 Determine the logic function of the ECL circuit in Figure 2.21.

2.13 What is the main advantage of I^{2}L gates.

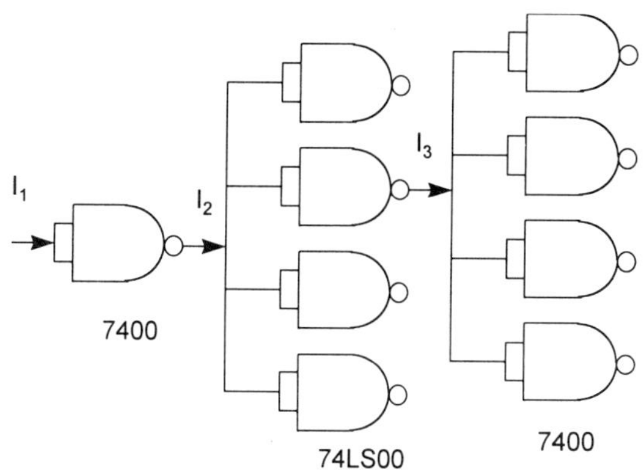

Figure 2.18 TTL circuit.

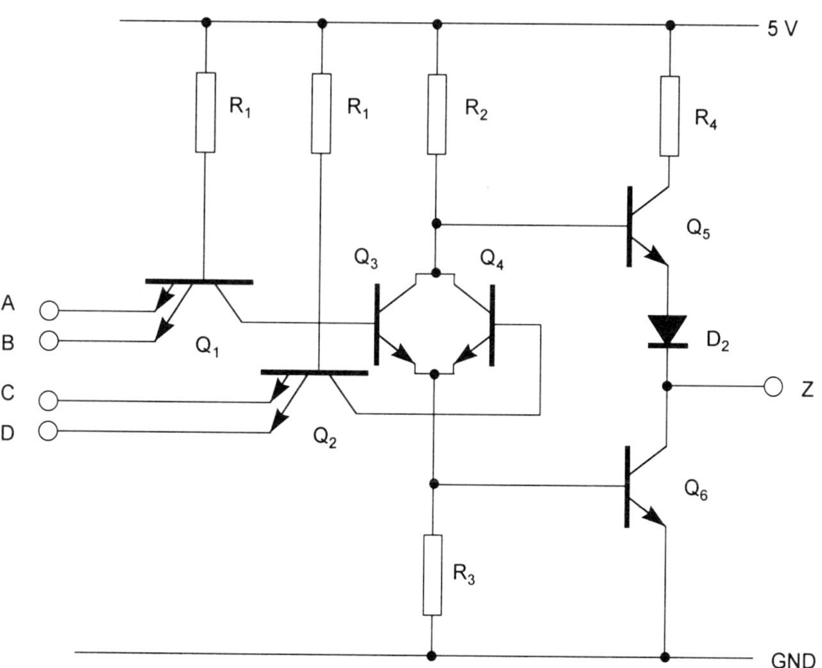

Figure 2.19 TTL gate.

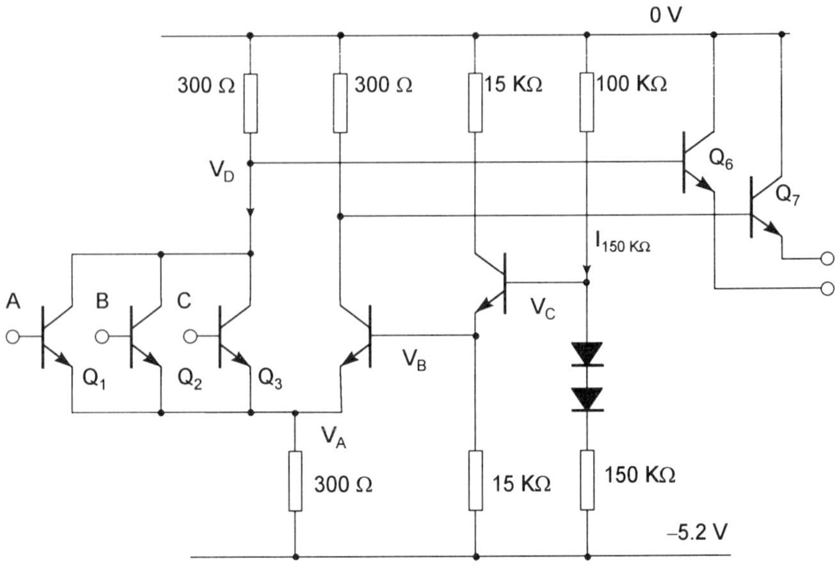

Figure 2.20 User-designed ECL gate.

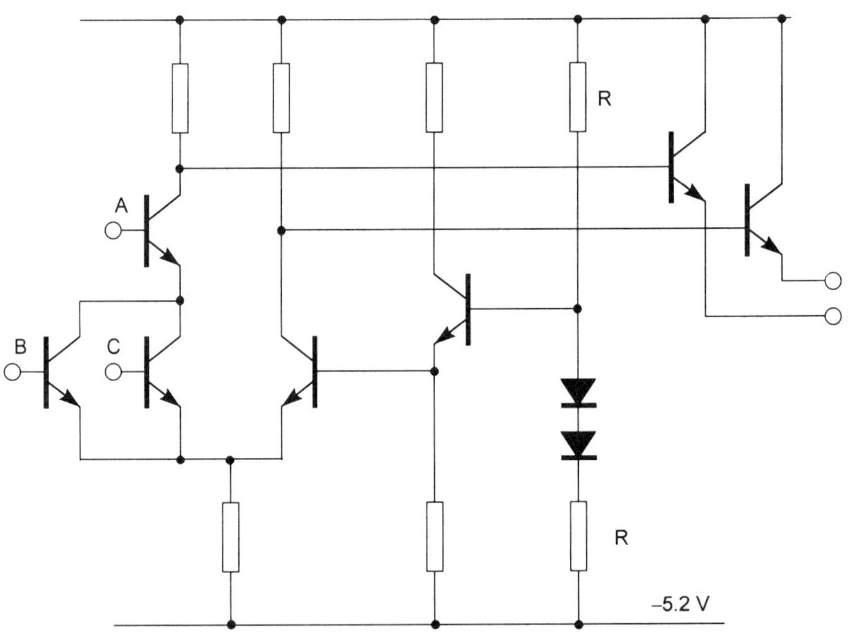

Figure 2.21 ECL gate.

MOS Design

3

3.1 Introduction

MOS devices have many advantages over bi-polar devices including the fact that they operate well as electrical switches and dissipate less power. Unfortunately, they are more susceptible to damage from electrostatic charges.

Another advantage of CMOS and NMOS design is that logic gates require much fewer transistors than TTL. There is also little no need for other passive elements, such as resistors and capacitors. It can be seen in Chapter 2 that a TTL inverter circuit required four transistors and many resistors. The bi-polar transistors are simple to produce but resistors tend to take up a relatively large amount of space. A comparable CMOS or NMOS design only uses two transistors. This allows NMOS and CMOS designs to contain extremely large densities of transistors. NMOS designs can have even larger densities than CMOS because NMOS logic normally uses fewer transistors for equivalent logic functions.

This chapter discusses logic design using CMOS and NMOS. Most large-scale designs now use either CMOS or NMOS. The great advantage of NMOS circuits is that a fabricated design can be scaled down to a smaller size without having to re-design the silicon layout.

With a MOS transistor the gate controls the current between the drain and the source. Unlike a bi-polar transistor, current can flow from the drain to the source or from the source to the drain. It can thus act as a bi-directional switch. With an npn bi-polar transistor current only flows between the collector and emitter and not vice-versa.

3.2 MOS devices

A MOSFET device can either be an n-channel MOSFET (an NMOS transistor) or a p-channel MOSFET (a PMOS transistor). Figure 3.1 shows the common symbols for NMOS and PMOS transistors. They are transconductance devices, where the gate-source voltage controls the drain-source current. In the case of logic design the transistor operates in saturation, that is, it is either ON or OFF. A large enough gate-source voltage results in a drain-source resistance which is almost zero. When the applied gate-source voltage is below a certain threshold then the drain-source resistance can be assumed to be infinite, as illustrated in Figure 3.2.

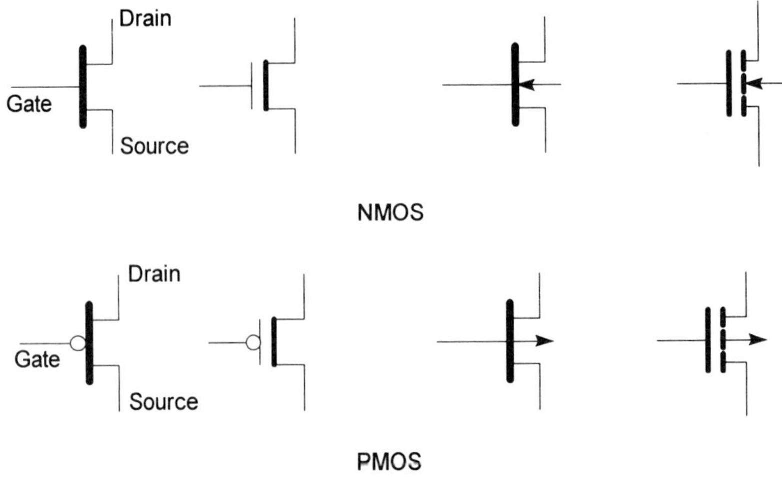

Figure 3.1 PMOS and NMOS symbols.

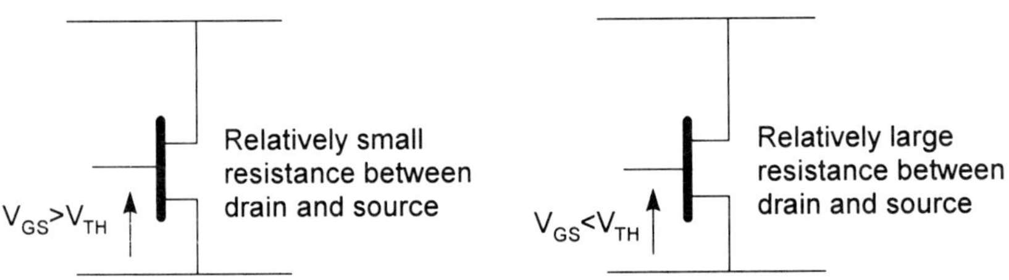

Figure 3.2 Approximation for NMOS transistor operation.

An important characteristic of MOS device is that the drain and source are interchangeable. This allows signals to pass from the drain to the source and vice versa.

3.3 CMOS logic gates

CMOS stands for complementary metal oxide semiconductor and is a popular type of MOS logic. A standard 4000 series of ICs are available as well as 74C that are pin-compatible with the TTL 74-series.

CMOS consists of both n-channel and p-channel MOS transistors (NMOS and PMOS) in a complementary connection. An NMOS transistor conducts when the gate-source voltage is higher than the gate-source threshold voltage. Thus an NMOS device conduct when there is a HIGH input on the gate and will not conduct for a LOW input. This is the opposite for a PMOS device, that is, a HIGH input causes it to stop conducting. Figures 3.3 and 3.4 show a CMOS two-input NOR gate and a two-input NAND gate.

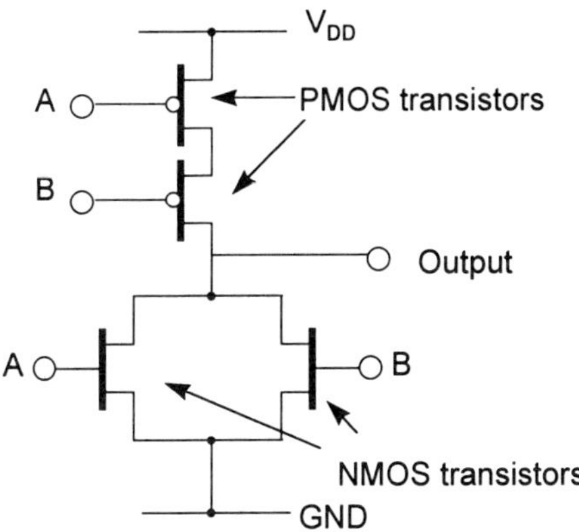

Figure 3.3 CMOS NOR gate.

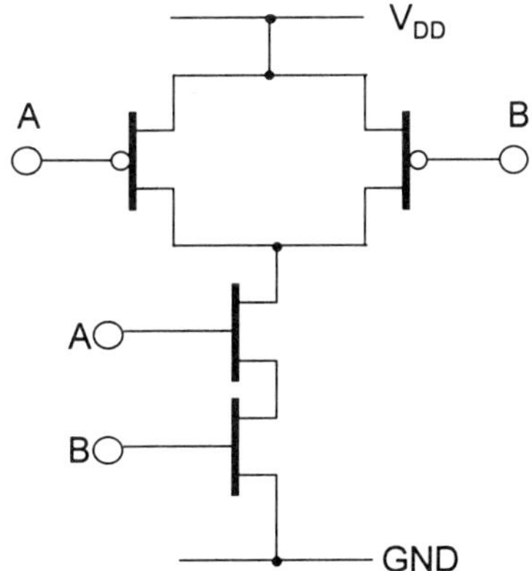

Figure 3.4 CMOS NAND gate.

CMOS speed is comparable to the 54L/74L series of TTL and several times slower than Schottky TTL. It has excellent noise immunity and its power dissipation is nearly zero. This lower power dissipation is due to the output being connected either to GND or to V_{CC}. The devices either have a very high resistance (minimum current flow through the device) or a very low resistance (minimum voltage drop across the device).

Unlike TTL, which is intolerant of large ranges of DC supply voltages, CMOS can operate over a supply range of +3 V to +15 V. Low power consumption and wide power supply ranges make CMOS ideal for battery operated equipment.

CMOS also interfaces well to TTL when operated with a +5 V power supply. The input impedance of CMOS is high enough so that TTL can drive it without any loss of fan-out capability in the LOW state. However, the HIGH level voltage of TTL is typically +3.5 V which is insufficient to drive CMOS reliably. A pull-up resistor connected from the output of the TTL drive to V_{CC}, as shown in Figure 3.5, increases the HIGH level to drive the CMOS device. The voltage drop across the 1 K resistor will be almost zero.

When CMOS uses supply voltages higher than +5V it is not possible to interface directly with TTL. In this situation a level translator circuit converts TTL levels to CMOS levels.

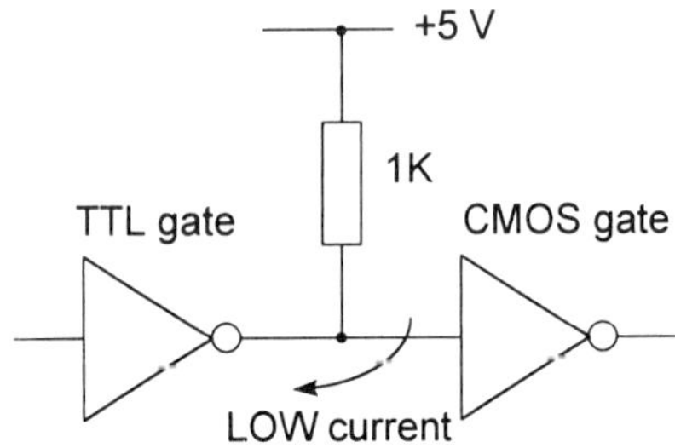

Figure 3.5 TTL to CMOS interface for 5V operation.

3.4 Designing CMOS logic

The upper PMOS block of a CMOS logic circuit provides a path to the HIGH level (V_{DD}) and the lower block provides a path to the LOW level (ground), as illustrated in Figure 3.6. The number of transistors used in a CMOS design can be significantly reduced by determining the required logic for the path to a HIGH level and also for the LOW level. This is done by splitting the Boolean function into a sum of products or the product of sums, then determing the switched path to the HIGH level and to the LOW level. The best way to show the technique is with an example.

First the upper block is designed:

if $\quad Z = \overline{A.B} + \overline{C}.D$

Next it is broken-up to either give a sum of products or a product of sums. The equation above is most suitable for conversion to a sum of products:

$$Z = \overline{A} + \overline{B} + \overline{C}.D$$

Figure 3.7 (a) shows how this function could be represented with PMOS transistors. The logic of the variables (A, B, C and D) has been changed because PMOS transistors operate with negative logic, that is, if the input is HIGH then the transistor is OFF. It can be seen that when A is LOW or if B is LOW or, if C is LOW and D is HIGH, then there is a path to the HIGH level, V_{DD}. Thus the output will be HIGH.

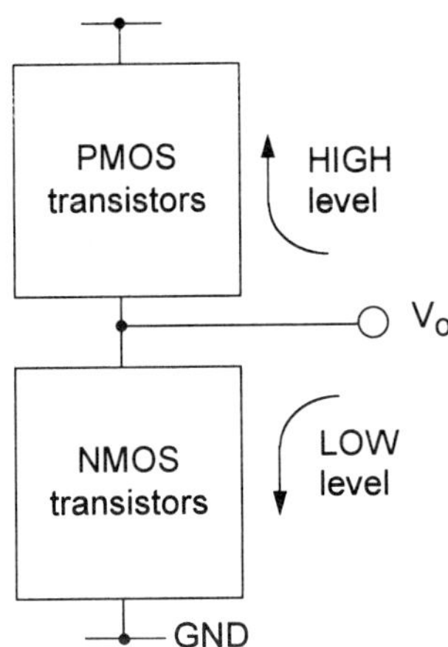

Figure 3.6 Provision of a LOW and HIGH level.

Next the path to the LOW level is determined by inverting the function, this gives:

$$\overline{Z} = \overline{\overline{A}.\overline{B} + \overline{C}.D}$$

Thus

$$\overline{Z} = (A.B).\overline{(\overline{C}.D)}$$
$$= (A.B).(C + \overline{D})$$

Figure 3.7 (b) shows how this function could be represented using NMOS transistors. In can be seen that when A is HIGH and if B is HIGH and if either C is HIGH or D is LOW then there is a path to the LOW level, GND. Thus the output will be LOW.

Figure 3.7 (a) provides a path to the HIGH level and Figure 3.7 (b) a path to a LOW level. These two circuits can be joined to produce the completed gate. This is shown in Figure 3.8. The inversion of the input D can be achieved with a simple inverter circuit. Thus the design requires eight transistors (and two transistors for the inverter).

If it had been designed using several stages of NAND, NOR and NOT gates it would have increased the propagation delay and would require many more transistors. Figure 3.8 shows one possible implementation which uses 18 transistors and five stages. It can thus be seen that the design in Figure 3.9 is an efficient one.

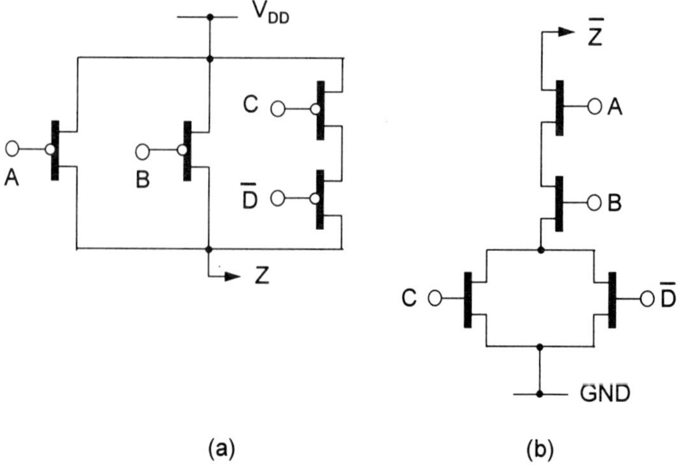

Figure 3.7 (a) Path to HIGH and (b) Path to LOW levels in the design example.

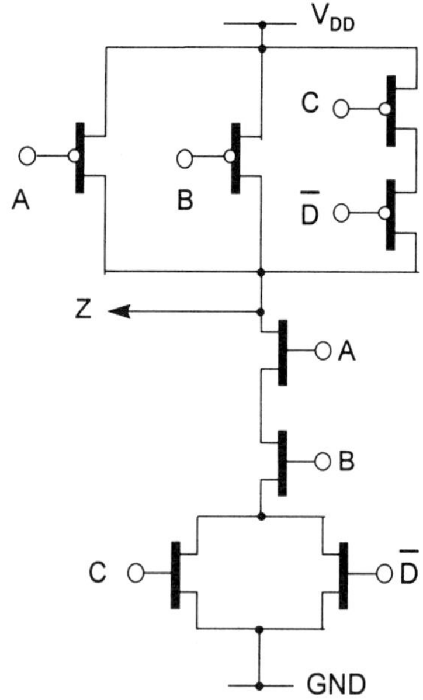

Figure 3.8 CMOS gate for $Z = \overline{A.B + \overline{C}.D}$

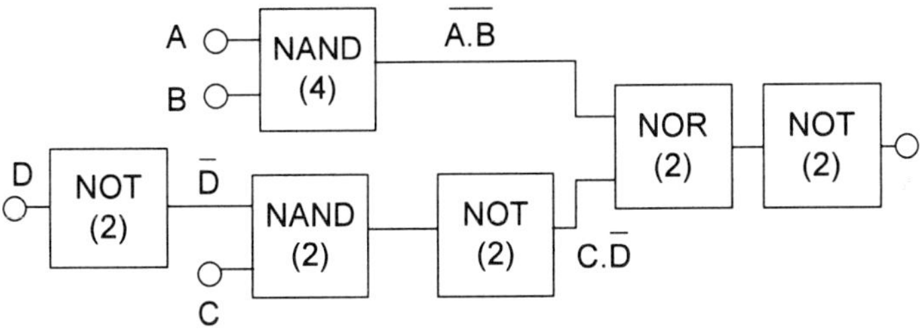

Figure 3.9 Alternative implementation of $Z = \overline{A.B} + \overline{C}.D$ using multiple stages.

3.5 PMOS logic gates

The first available high-density MOS circuits used PMOS which use enhanced-mode p-channel MOS transistors to form basic logic blocks. Figure 3.10 shows a PMOS gate for a positive logic NOR function.

The supply voltage V_{GG} is a negative voltage, and V_{CC} is a positive voltage, or possibly ground (0 V). An enhanced-mode MOS transistor conducts when the applied gate-source voltage is zero. Thus the enhanced-mode transistor in Figure 3.10 always conducts because its gate connects to its source. This causes a constant drain-source resistance which limits the current in the circuit. The usage of the depletion mode transistor and enhanced-mode transistors will be discussed in more detail in Chapter 4.

For a LOW (V_{GG}) applied to either input A or B causes Q_1 or Q_2 to be ON. The output is then pulled up towards GND, which represents a HIGH. When a HIGH voltage (GND) is applied to inputs A and B, transistors Q_1 or Q_2 are turned OFF. This causes the output to be pulled down to a LOW (near V_{GG}). Thus a LOW output occurs when both inputs are HIGH, and a HIGH output occurs only when either input is LOW. The actual voltage level for a LOW depends on the resistance of transistor Q_3 and the resistance of the pull-up transistors, Q_1 and Q_2.

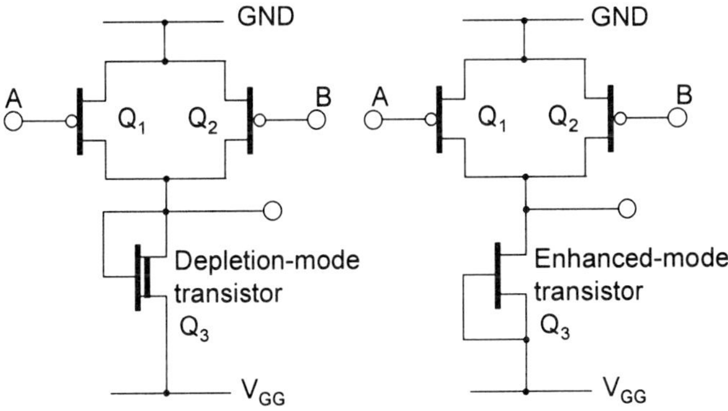

Figure 3.10 Basic PMOS NOR gate.

3.6 NMOS logic gates

NMOS logic devices use n-channel MOS transistors. They were developed as MOS processing improved and now are used in most memories and microprocessors. Figures 3.11 and 3.12 show a NAND and NOR gate.

With the NAND circuit, Q_3 acts as a current limiter. When a LOW (V_{GG} or ground) is applied to one or both inputs, then at least one of the transistors (Q_1 or Q_2) is OFF which makes the output HIGH (near V_{CC}). When a HIGH (V_{CC}) is applied to both inputs then both Q_1 and Q_2 conduct and the output is LOW. The actual output voltage level depends on the drain-source resistance of the three transistors.

With the NOR gate, Q_3 again acts as a resistor. A HIGH on either input turns one of the input transistors (Q_1 or Q_2) ON, thus making the output LOW. When both inputs are LOW, both transistors are off and the output is pulled up to a HIGH level.

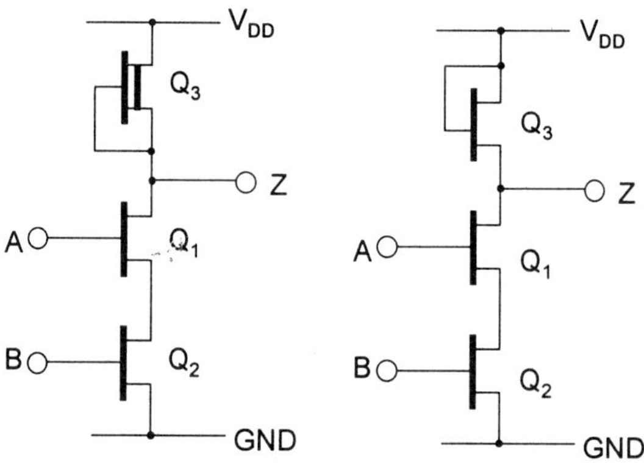

Figure 3.11 NMOS NAND gate.

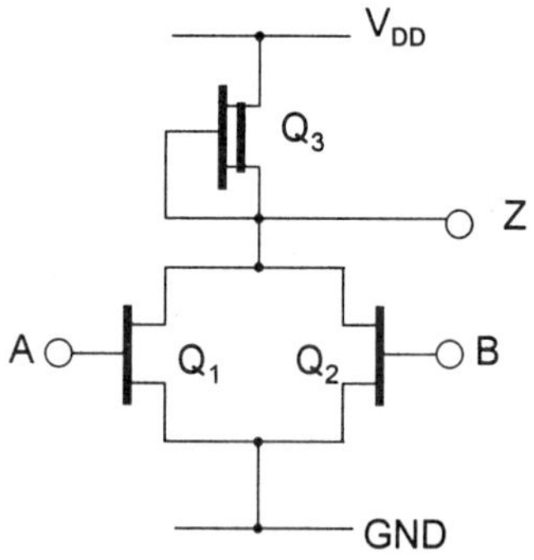

Figure 3.12 NMOS NOR gate.

3.7 CMOS .vs. TTL

CMOS has a much higher fan-out than TTL because of the lower input and output currents. The only limiting factor is switching speed because the more output loads there are the greater the loading capacitance. A standard TTL gate has a maximum fan-out of 10 UL, whereas a typical CMOS gate has a maximum fan-out exceeding 50 UL. Table 3.1 shows typical comparisons of the two families.

Table 3.1 Comparison between TTL and CMOS.

	74	*74L*	*74LS*	*CMOS (5V)*	*CMOS (10V)*
Propagation delay	10 ns	33 ns	9.5 ns	40 ns	20 ns
Quiescent power	10 mW			10 nW	10 nW
Noise immunity	1 V	1 V	0.8 V	2 V	4 V
Fan-out	10	10	20	50+	50+
Clock frequency	35 MHz	3 MHz	45 MHz	8 MHz	16 MHz

3.8 Noise immunity

Noise immunity is a gates ability to withstand fluctuations of the voltage levels within its operating range. The DC noise margin of a logic gate is the measure of its noise immunity. For example in TTL a '1' is defined as any voltage above 4 V; a voltage below this level may be a '1' or a '0'.

3.9 Bi-CMOS

Bi-polar devices have low propagation delays and good output current drive, whereas CMOS have high impedance inputs and low power dissipation. These two types of devices are combined in Bi-CMOS. Bi-polar transistors provide the output stage and CMOS the input and intermediate stages. Figure 3.13 shows a Bi-CMOS inverter.

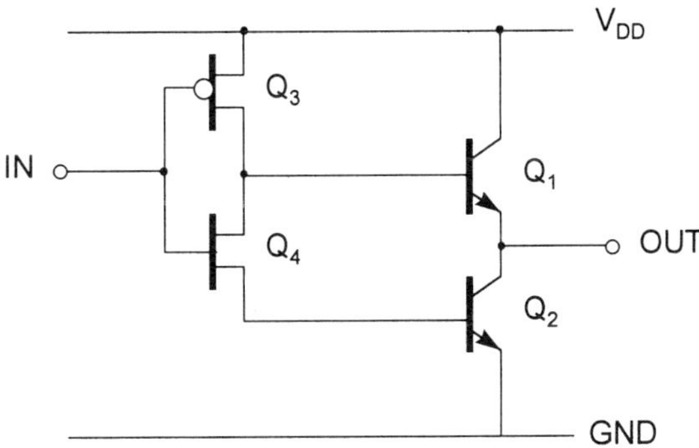

Figure 3.13 Bi-CMOS inverter.

A bi-polar push-pull circuit on the output isolates the MOS transistors from the load and also provides driving capabilities. When the input is HIGH, the pull-down MOS transistor (Q_4) short-circuits the base-collector junction of Q_2 and thus converts it into a diode. The pull-up MOS transistor is OFF and Q_1 has no base drive. The output will be low as Q_2 conducts and Q_1 does not, the equivalent loading is shown in Figure 3.14.

For a LOW input the reverse occurs. The transistor Q_3 short-circuits the base-collector junction of Q_1 makes the output HIGH, while Q_2 is OFF as it has no bias current, as illustrated in Figure 3.15.

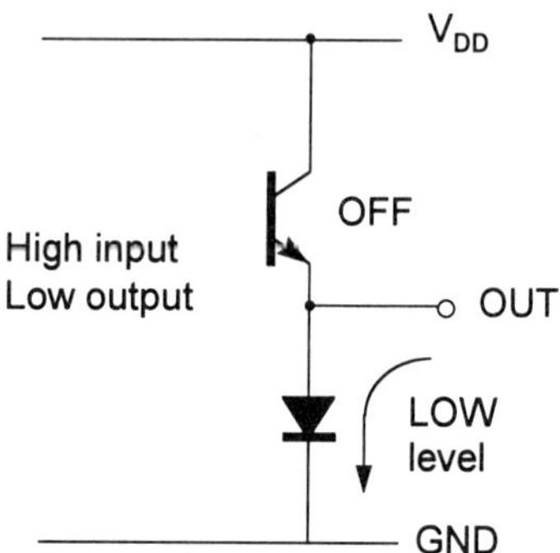

Figure 3.14 Output equivalent circuit with a HIGH input.

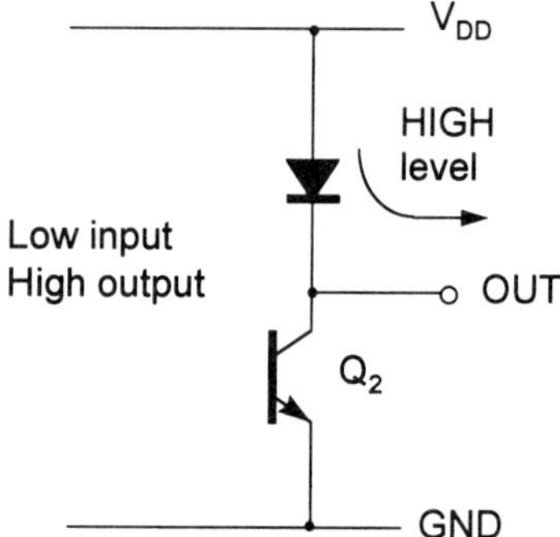

Figure 3.15 Output equivalent circuit with a LOW input.

3.10 MOS precautions

The most important precautions are:

1. That MOS devices are always shipped/stored in conductive foam to prevent charge build-up. When removed from the foam the pins they must not be touched by human hands.

2. The pins should be placed on a grounded surface, such as a metal plate, when removed from the protective material.
3. All tools, test equipment and metal workbenches should be earth-grounded. Anyone working with MOS devices should wear an earth strap.
4. MOS devices should not be inserted into sockets or PCBs with the power on. All unused inputs must be connected to V_{SS} or V_{DD}. Unused pins should not be left unconnected as they can acquire static charge and "float" to unpredictable levels.

3.11 Exercises

3.1 Explain why CMOS has a higher fan-out than TTL gates.

3.2 What is the main advantage of using NMOS logic over PMOS?

3.3 Explain why MOS logic technology allows higher transistor densities than bi-polar devices.

3.4 Explain the operation of a Bi-CMOS inverter circuit and determine typical output voltages for a HIGH and LOW when using +5 V and 0 V supply.

3.5 Determine the function of the gates given in Figure 3.16.

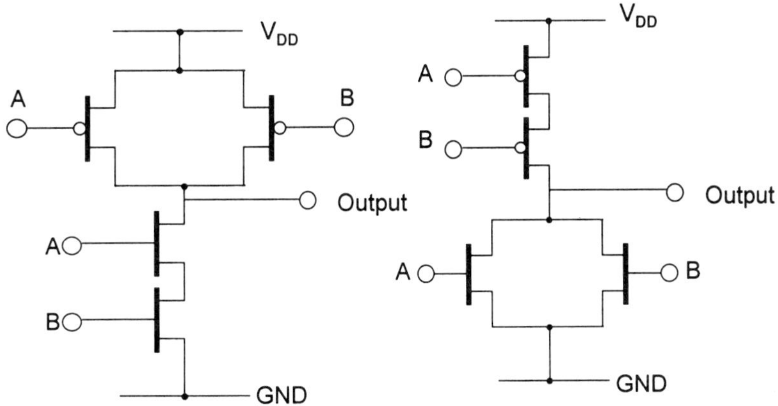

Figure 3.16 CMOS circuits.

3.6 Contrast TTL and CMOS outlining characteristics such as speed of operation, fan-out, noise immunity and power dissipation.

3.7 Design a CMOS gate using the minimum number of transistors, for the following Boolean equations:

(a) $Z = \overline{(A.B) + (C.D)}$

(b) $Z = \overline{A.(E + F) + B.C}$

Determine the number of required transistors and contrast this with a design using NAND, NOR and NOT stages.

3.8 Design a CMOS circuit, with a minimal amount of transistors to implement the following Boolean equation:

$$Z = \overline{[(A + B) \cdot (C + (A \cdot D))]}$$

MOS Device Modelling

4

4.1 Introduction

The two types of MOS device, n-channel MOS (NMOS) and p-channel MOS (PMOS) differ in the type of diffusion they use. Figure 4.1 shows a cross section of an enhanced-mode NMOS device and Figure 4.3 shows a depletion-mode device. An enhanced-mode NMOS transistor conducts when the gate-source voltage is greater than a positive threshold voltage. Whereas, a depletion-mode NMOS transistor conducts when the gate-source voltage is zero and requires a negative gate-source to stop it conducting. Figures 4.2 and 4.4 show the voltage-current characteristics for the two types.

NMOS devices are formed in a p-type substrate of moderate doping level. The source and drain regions have an n-type impurity concentration. This creates depletion regions which extend mainly in the more lightly doped p-regions, as illustrated in Figure 4.1.

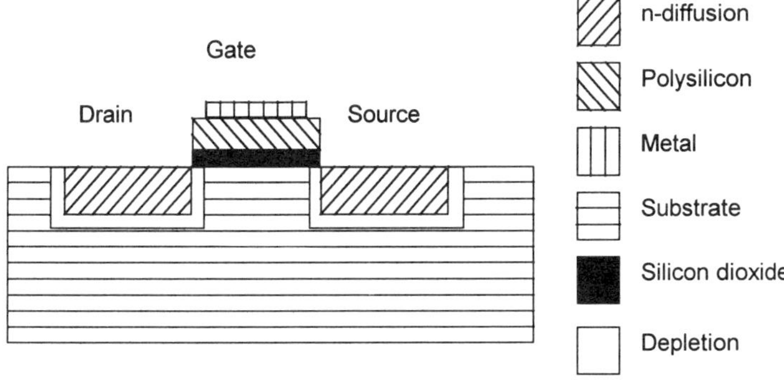

Figure 4.1 NMOS enhanced-mode transistor.

A layer of silicon dioxide (SiO_2) isolates the gate from the substrate and thus no current flows between the gate and source (or drain), unless the applied voltage is large enough to break through the isolation. MOS transistors are transconductance devices where a gate-source voltage causes a current to flow from the drain to the source. This is because the applied gate-source voltage produces an electric field which causes a depletion region in the substrate. For an enhanced-mode NMOS transistor a positive gate-source voltage produces a forward-biased gate-source junction. This helps the flow of current between the drain and source. No conduction occurs until the gate-source voltage is greater than a certain threshold voltage.

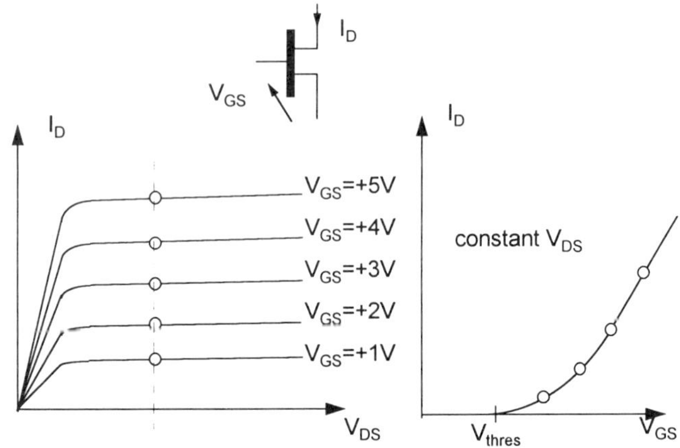

Figure 4.2 NMOS enhanced-mode V-I characteristics.

A depletion-mode transistor differs from an enhanced-mode transistor in that it has an added n- or p-type implant below the gate. In an NMOS transistor this is a lightly doped n-type diffusion and in a PMOS transistor it is p-type. This implant causes a current flow between the drain and source even when there is no applied gate-source voltage. As with an enhanced-mode transistor, an increasing positive gate-source voltage causes increasing drain-source current. This conduction only stops when the applied gate-source voltage is less than a negative threshold voltage.

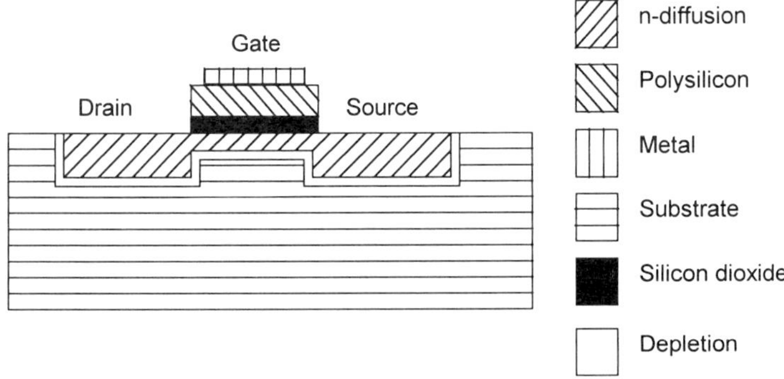

Figure 4.3 NMOS depletion-mode transistor.

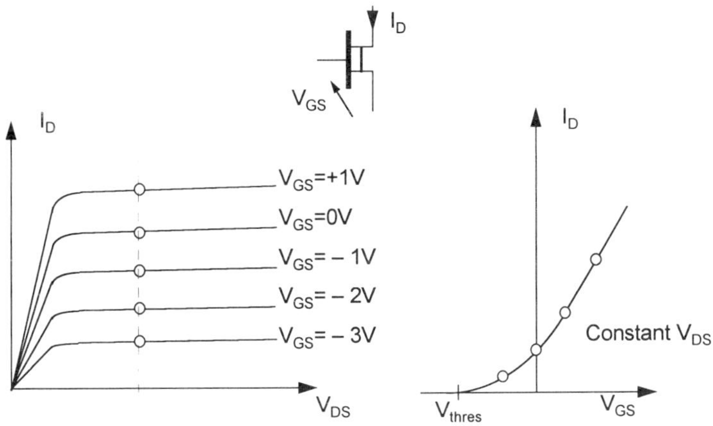

Figure 4.4 NMOS depletion-mode V-I characteristics.

4.2 MOS characteristics

The MOS transistor is a transconductance device, that is, a change in the gate-source voltage causes a change in the drain-source current. No conduction occurs in an enhanced-mode NMOS device until the gate-source voltage reaches a threshold voltage. Typically, this threshold is around 1 V for a 5 V supply voltage. Appendix A contains equations which define the relationship between the gate-source voltage and the drain current. Such complex equations are seldom required when designing logic gates. Normally a few approximations are made, these are:

- the ON drain-source channel resistance (R_{DS}) varies directly with the length and indirectly with the width of the gate;
- the OFF drain-source channel resistance is infinite.

Figure 4.5 shows two MOS transistors. The width and length of each gate defines their characteristics. As has been stated above, the resistance of the channel varies directly with the length of the gate and inversely with the width of the gate. Thus the transistor on the left-hand side has a greater channel resistance than the transistor on the right-hand side for the same applied gate source voltage.

4.3 Gate capacitance

The gate capacitance of a MOS device varies with the width and length of the channel. A layer of polysilicon sits above the substrate and is insulated by a layer of silicon dioxide. This arrangement approximates to parallel-plate capacitor and the approximate capacitance will be thus:

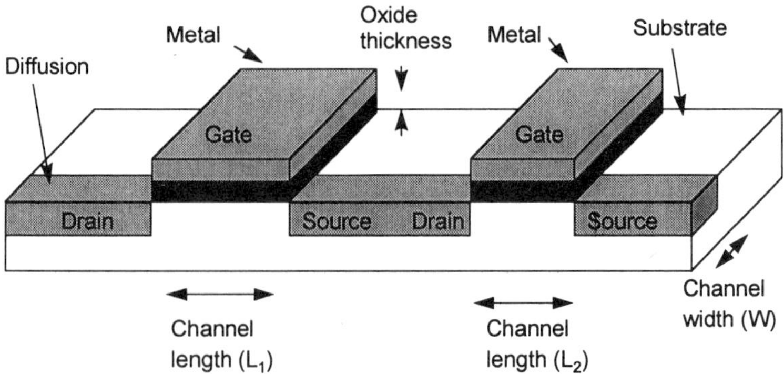

Figure 4.5 Structure of an MOS transistor.

$$C = \frac{\varepsilon . A}{d} \quad \text{Farads}$$

where A is the area of the plates and d the distance between the plates. Figure 4.5 shows how the gate can be approximately to a parallel-plate capacitor separated by a silicon dioxide layer.

The gate-source capacitance can thus be found by:

$$C_G = \frac{\varepsilon}{T_{ox}}(WL) \quad \text{Farads}$$

where T_{ox} is the thickness of the oxide layer. Thus the greater the width and length of the channel (or the gate), then the greater the input capacitance. An increased capacitance reduces the speed of the gate and thus increases the propagation delay in the circuit.

4.4 MOS devices as a switch

An enhanced-mode MOS transistor, in the linear region, operates as a voltage controlled resistor. When it is used at its operation limits it operates approximately as an electrical switch. The ON resistance is typically less than 100 Ω and the OFF resistance is normally greater than 100 MΩ.

When the gate-source voltage is greater than the threshold voltage then a current flows between the drain and the source. The drain-source voltage drop depends upon the channel resistance and will be given by $V_{DS}=I_{DS}.R_{CH}$. If the load resistance is much greater than the channel resistance then the voltage drop across the drain source will be small.

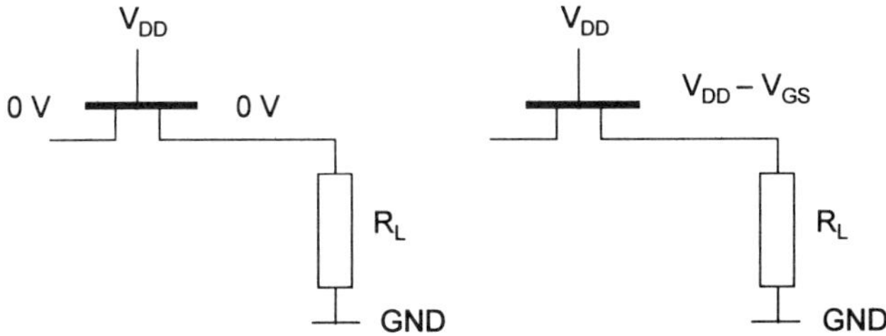

Figure 4.6 Signal reduction due to the pass transistor for an NMOS transistor.

A MOS transistor is not a perfect switch because it has a finite resistance in the ON state and because it requires a finite gate-source voltage to be able to turn the switch ON. To minimize the voltage drop across the MOS transistor channel, the load impedance must be much larger that the ON resistance of the FET. When this is the case a low applied voltage causes the signal to pass to the load with little degradation. Unfortunately, when the input voltage is equal to the applied gate voltage then the output voltage will be $V_{DD}-V_{GS}$, as illustrated in Figure 4.6. Thus the maximum voltage that can be output will be $V_{DD}-V_{TH}$, since beyond this value the gate-source voltage drops below threshold and the transistor turns OFF.

A FET switch operates well if the load resistance is greater than the channel resistance and can pass signals in either direction, that is, from drain to source or from source to drain. Thus, it acts as a truly bilateral switch.

Figure 4.7 shows a basic NMOS inverter circuit with a pull-up resistor which limits currents when the transistor is ON. For example, a 40 kΩ pull-up resistor gives a drain-source current of 125 μA for a 5 V supply (5/40 000).

When the input voltage (V_{IN}) is less than the threshold voltage then the transistor will be OFF and the output will be HIGH, or V_{DD}. For an input voltage greater than the threshold voltage then the transistor conducts. The transistor eventually saturates with a large enough input voltage. When $V_{IN}=V_{DD}$ then the output voltage will be:

$$V_{LOW} = V_{DD} \frac{R_{CH}}{R_{CH} + R_L}$$

where R_{CH} is the channel resistance when $V_{GS}=V_{DD}$.

When choosing the pull-up resistor there is a trade off between power dissipation and speed. The lower the pull-up resistance then the faster the circuit responds (as the RC time constant is lower) but this leads to a higher current when the output is LOW. This leads to higher power dissipation. Note that there will be very little current flow and associated power dissipation when the output is a HIGH level.

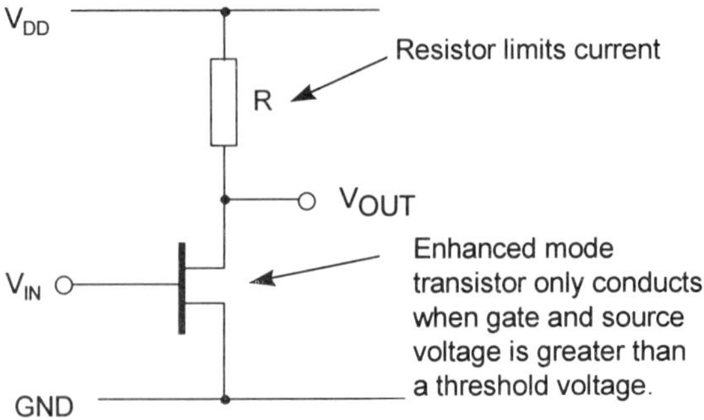

Figure 4.7 NMOS inverter.

4.5 NMOS inverter with a depletion load

Resistors are difficult to fabricate on silicon as they require a relatively large area. An improved method is to use a depletion-mode transistor as a pull-up resistor. Figure 4.8 shows an inverter circuit with a depletion-mode transistor. As previously mentioned, the depletion-mode transistor conducts when the gate-source voltage is zero.

The resistance of a drain-source channel, with a length L and width W, is:

$$R = \frac{L}{W} \times Sheet\ \ resistivity$$

where the sheet resistivity is defined as the resistance of a unit square of the material. When V_{IN} is LOW, the pull-down transistor is OFF and the equivalent electrical circuit is a single resistance of R_{PU} connected to V_{DD}. R_{PU} is the resistance of the pull-up transistor, as illustrated in Figure 4.9 (a).

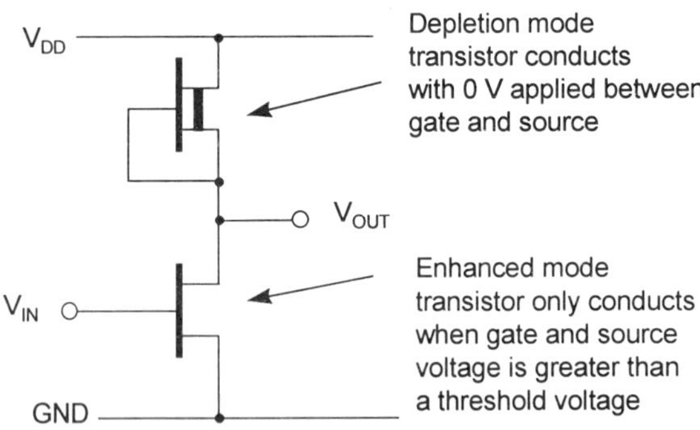

Figure 4.8 NMOS inverter with pull-up load.

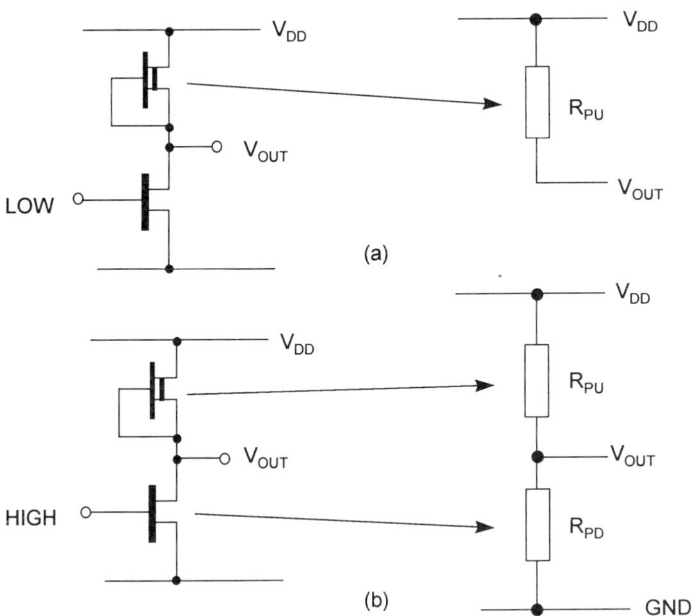

Figure 4.9 (a) V_{IN} is LOW (b) V_{IN} is HIGH.

When V_{IN} is HIGH the inverter is equivalent to a voltage divider, as shown in Figure 4.9 (b), with R_{PU} connected in series with R_{PD}. The output voltage is the voltage across R_{PD} and is given by:

$$V_{OUT} = V_{DD} \frac{R_{PD}}{R_{PD} + R_{PU}}$$

By assuming that the threshold is one-fifth of the supply voltage then it can be shown that:

$$R_{PU} \geq 4R_{PD}$$

Since R_{PU} is directly proportional to $(L/W)_{PU}$ and R_{PD} to $(L/W)_{PD}$, gives:

$$\left(\frac{L}{W}\right)_{PU} \geq 4.\left(\frac{L}{W}\right)_{PD}$$

The inverter ratio, k, is the ratio of the length to width ratio of the pull-up transistor to the length to width ratio of the pull-down transistor. Thus the inverter ratio of the inverter must be greater that or equal to four. Thus k is:

$$k \geq \frac{\left(\frac{L}{W}\right)_{PU}}{\left(\frac{L}{W}\right)_{PD}}$$

4.6 NMOS NAND and NOR gates

Figure 4.10 shows an NMOS inverter gate. It was seen that assuming a threshold voltage of one-fifth of the supply voltage that the inverter ratio (k) must be at least 4. The length to width ratio of the pull-up transistor is four times greater than the pull-down transistor. This makes the ON resistance of the pull-up transistor four times greater the pull-down transistor. This provides an output LOW which is less than the threshold voltage of the next stage. Thus if the L:W ratio of the pull-up transistor is 2:1 then the pull-down transistor must be ratioed 1:2. If the pull-up transistor was ratioed 4:1 then the pull-down transistor would be ratioed 1:1.

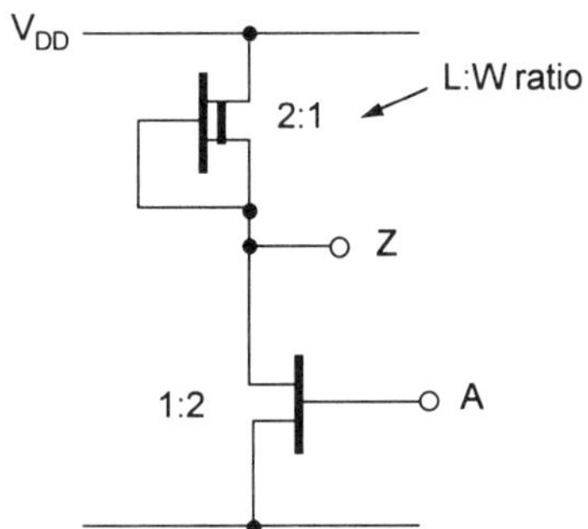

Figure 4.10 Inverter gate.

Figure 4.11 shows a basic NMOS NAND depletion-mode pull-up gate. In this case the inverter ratio k should be at least 8. When the output is a LOW then the pull-down resistance will be $R_{PD}+R_{PD}$, thus the LOW output voltage level for the NAND gate will be:

$$V_{LOW} = V_{DD} \frac{R_{PD} + R_{PD}}{R_{PD} + R_{PD} + R_{PU}}$$

$$0.2 = \frac{R_{PD} + R_{PD}}{R_{PD} + R_{PD} + R_{PU}}$$

$$0.4R_{PD} + 0.2R_{PU} = 2R_{PD}$$

$$0.2R_{PU} = 1.6R_{PD}$$

Thus the inverter ratio for a 2-input NAND gate is:

$$k = \frac{R_{PU}}{R_{PD}} = \frac{1.6}{0.2} = 8$$

Figure 4.12 shows a basic 2-input NOR gate. It can be shown that the inverter ratio must be at least 4.

Each of the transistors in Figure 4.11 and 4.12 are identified by their length to width (L:W) ratio. This ratio is also known as the aspect ratio of the transistor. For example the depletion-mode transistor in Figure 4.11 has an aspect ratio of 4 thus its channel length is four times longer than its width.

The aspect ratio of the pull-up transistor of the gates determines the gate power dissipation. A maximum power dissipation of a NAND gate occurs when all the inputs are HIGH, whereas for a NOR the maximum power dissipation occurs when any of the inputs is a HIGH.

Speed is also a critical factor, and the rise-times of the output voltage increases as the NAND gate pull-up inverter ratio increases. If the pull-down aspect ratios are fixed, the pull-up aspect ratio and the rise-time of the NAND gate are proportional to the number of inputs, whereas the NOR gate has both a fixed pull-up aspect ratio and a constant rise-time. In general, NMOS NAND gates are slower than NMOS NOR gates with the same number of inputs and power dissipation.

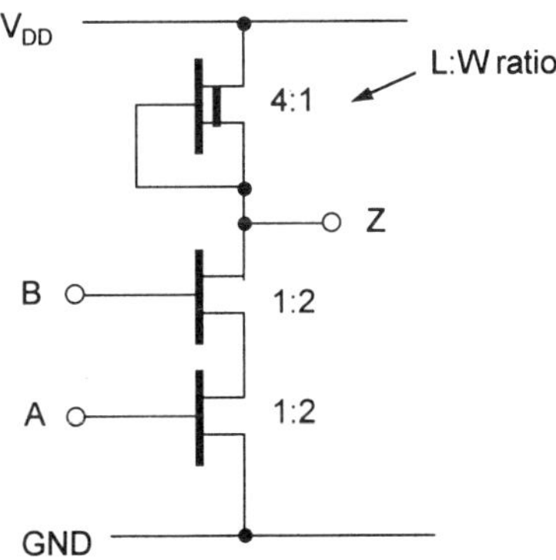

Figure 4.11 NAND gate.

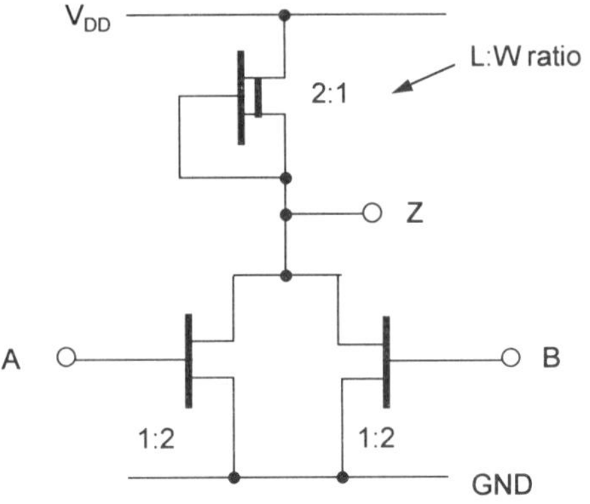

Figure 4.12 NOR gate.

4.7 Exercises

4.1 A transistor has an oxide layer of 700 Å (70 nm). The permitivity of gate insulator is $3.\varepsilon_0$ (ε_0=8.85x10^{-14} Farads.cm^{-1}). If the width and length of the gate is 5×10^{-4} cm, determine the gate capacitance.

4.2 An NMOS transistor has an SiO_2 thickness of 70 nm, a gate width of 3 μm and gate length of 1.5 μm. Determine the gate capacitance, in pF.

4.3 Show that, for the circuit given in Figure 4.13, the drain-source resistance for the saturated transistor should be four times the pull-up resistance. Assume that the threshold voltage of the transistor is one-fifth of V_{DD}.

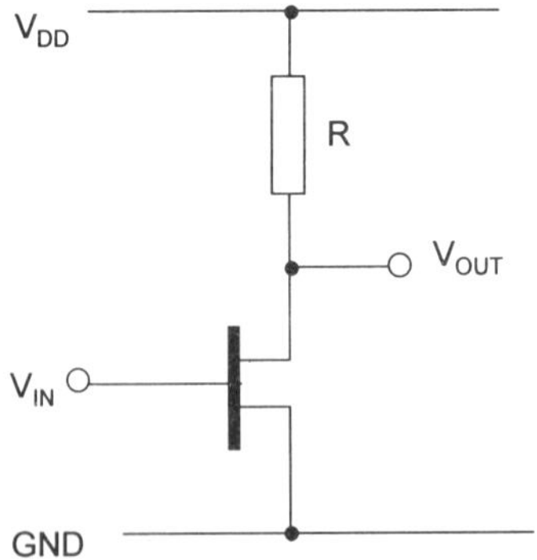

Figure 4.13 NMOS circuit.

4.4 Determine the inverter ratio, k, for the following gates:

(a) inverter;
(b) 2-input NAND
(c) 2-input NOR.

Explain how the inverter ratio changes for NAND and NOR gates with any number of inputs. Determine k when for a 3-input NAND and a 3-input NOR gate.

4.6 Explain how rise-time and power dissipation vary with differing aspect ratios for the NAND and NOR gates.

NMOS Circuit Design 5

5.1 Introduction

Chapter 4 analyzed NMOS and CMOS logic gates. This and the next chapter cover the design and full custom layout of NMOS and CMOS transistor circuits. It also provides an understanding of how digital circuits are laid out on silicon and how they interconnect. The main advantages of NMOS circuits are that the transistor densities are extremely high and they produce scaleable designs that can be down-sized (or up-sized) for a different fabrication process (that is, a different transistor size). The Intel Pentium processor contains over 3.2 million NMOS transistors and the P6 should have over 6 million. These transistor densities are only possible with smaller minimum sizes and NMOS design techniques. One problem associated with NMOS circuits is that they dissipate more power than CMOS circuits. This higher dissipation is mainly due to the active pull-up load.

Semi-custom design involves designing structures from the silicon upwards (that is, bottom-up design). Typical structures which are designed using bottom-up designs are memory arrays, logic gates (such as basic NAND, NOR or NOT gates), buffer circuits, regular array structures, and so on. This type of design involves full-custom or semi-custom design techniques.

5.2 NMOS fabrication process

Diffusing n-type doping into a p-type substrate produces NMOS transistors. The n-diffusion region creates the drain and source of the transistor and polysilicon creates the gate. There are six main materials used in the NMOS design process, these are:

- polysilicon;
- metal;
- ion implant for depletion mode transistors;
- silicon dioxide;
- n-type diffusion;
- p-type substrate.

The NMOS design process uses a number of masks for each of the layers. Each of these masks is colour coded, such as blue for metal, red for polysilicon, green for n-diffusion. Appendix B discusses the fabrication process. The six basic masks involved in fabrication are:

- *mask 1* the SiO_2 layer is etched to expose paths in the diffusion layer or source, drain or gate area of transistors (sometimes named a diffusion mask);
- *mask 2* ion implant for depletion mode transistors;
- *mask 3* deposit polysilicon layer;
- *mask 4* oxide layer and etch for contact cuts;
- *mask 5* deposit metal layer;
- *mask 6* overglassing process.

5.3 Stick diagrams

Several masks define the layout of transistor and connections. Each mask uses a different colour to identify its function. Table 5.1 outlines the representation of the layers as either a coloured mask or stick diagram. In NMOS design the source and drain are green (brown in PMOS) and the gate is red.

Table 5.1 Layer representations.

Function	*Stick diagram*	*Colour code*	*Mask*
Polysilicon	– – – – – –	RED	See Figure 5.1
Diffusion	———	GREEN (n-diffusion) BROWN (p-diffusion)	See Figure 5.1
Metal	═══	BLUE	See Figure 5.1
Contact or Via	× ●	BLACK	■
Implant	⌜ ⌝ ⌞ ⌟ (dashed box)	YELLOW	(dashed box)

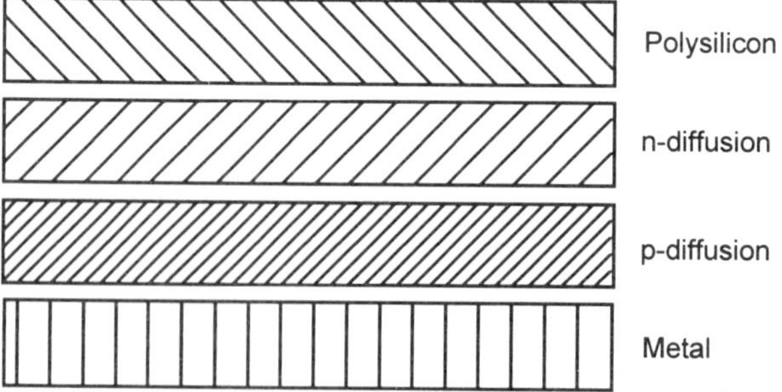

Figure 5.1 Mask representations for polysilicon, n-diffusion, p-diffusion and metal.

A three-level MOS, single-metalization process is analogous to a three-layered printed-circuit board (PCB) consisting of three levels of conducting material. Silicon dioxide (SiO_2) isolates each layer from the layers above and below it, as shown in Figure 5.2. A metal via connect one layer to another and a contact cut connects polysilicon to diffusion.

The three-level NMOS and PMOS structure expands to two or more metal or polysilicon layers by adding more masks and by assigning different colours to the other metal and/or polysilicon layers. This gives four or more levels. A second metal layer uses purple or a different shade of blue, a second polysilicon mask uses a different colour or shade of red.

Metal can cross polysilicon or diffused layers with no functional effect other than to produce a parasitic capacitance. Thus, blue conductive layers can cross red and green conductive layers. Metal connects to another layer using vias (black). Polysilicon crossing a diffused area creates a MOS transistor. Then whenever a red line crosses a green line it creates an n-channel transistor (NMOS), as illustrated in Figure 5.3. A red line crossing brown creates a p-channel transistor (PMOS).

Figure 5.4 shows the representation of a depletion-mode NMOS transistor. A hatched line or yellow colouring defines the implant region.

The gate (or channel) width (W) and length (L) defines the characteristics of the MOS transistor. The ON resistance of the transistor varies directly with the length and indirectly with the width. The dimensions W and L define the minimum size of the transistor. These absolute sizes are dependent on the process, a typical technology is W=3 μm (3 microns) and L=1.5 μm (1.5 microns). L defines the channel length from the source to drain and W the channel width. In summary:

1. Polysilicon crossing n-diffusion creates an NMOS transistor.
2. Metal crossing polysilicon or n-diffusion has no effect (other than parasitic capacitance).

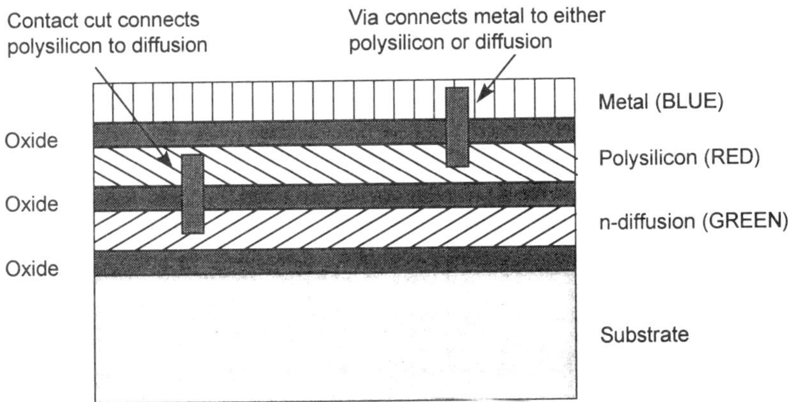

Figure 5.2 Three-level MOS with one layer of metal.

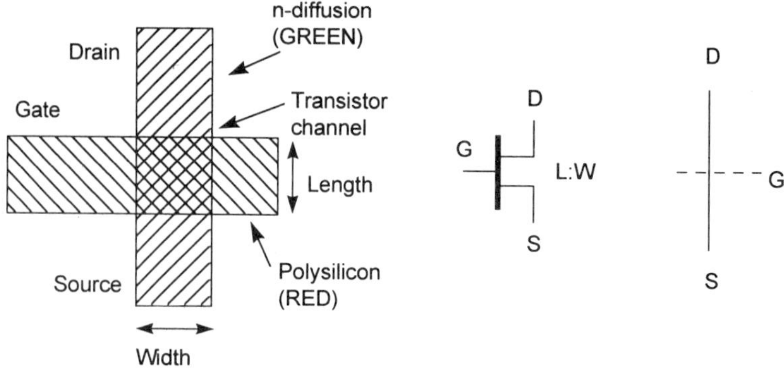

Figure 5.3 Layout of an enhanced-mode NMOS transistor.

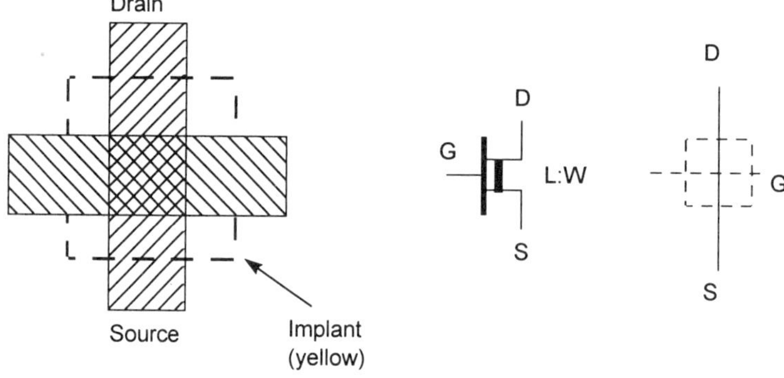

Figure 5.4 Layout of a depletion-mode NMOS transistor.

5.4 Lambda design rules

A great advantage of NMOS design is that the scaling the dimensions of the transistors has little effect on their logic operation. The ratio of the gate length to its width is critical to the operation. For this reason all transistors are scaled relative to a size λ (lambda). At the time of fabrication this is assigned an absolute size, such as 3 μm.

Mead and Conway developed design rules which define the minimum sizes of transistors, minimum separation distances, and so on. Table 5.2 outlines these. The rules are analogous to the rules applied in PCB design, such as minimum track size, separation between tracks, and so on.

Other rules can be applied to the design. The most prevalent is the 'micron-based' rule set which will be discussed later.

Table 5.2 Minimum lambda rules.

Description	*Minimum size*
Thickness of polysilicon lines	2λ
Distance between polysilicon lines	2λ
Thickness of n-diffusion lines	2λ
Distance between n-diffusion lines	3λ
Distance between polysilicon and n-diffusion	λ
Minimum thickness of metal	3λ
Distance between metal	3λ

A summary of the lambda design rules are:

1. The minimum width of a transistor is 2λ and the minimum length is 2λ. The minimum sizes of an enhanced- and depletion-mode NMOS transistors are given in Figure 5.5.
2. The implant region of a depletion-mode transistor is at least 6λ in both dimensions and must extend at least 2λ on either side of the polysilicon polygon. The separation between the implant region and another transistor should be 2λ.
3. A via (or contact cut) and a transistor must be separated by at least 2λ.
4. Minimum thickness of metal is 3λ.
5. Minimum thickness of polysilicon and diffusion lines are 2λ.
6. Minimum separation between n-diffusion and polysilicon lines is λ.
7. Minimum separation between polysilicon lines is 2λ.
8. Minimum separation between diffusion lines is 3λ.

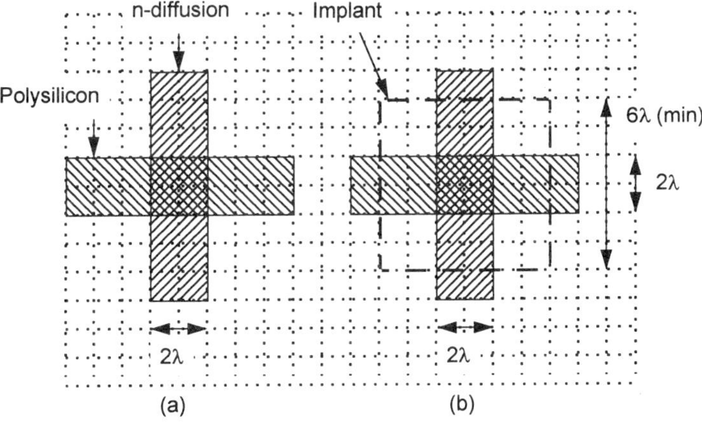

Figure 5.5 Minimum sizes for (a) enhanced-mode NMOS transistor and (b) depletion-mode NMOS transistor.

5.5 Contact cuts

Contacts between polysilicon and n-diffusion are typically required for depletion-load transistors. They can be achieved either by:

- adding a buried contact, which connects the two layers directly;
- connecting the polysilicon to metal then connecting the metal to diffusion.

The buried contact method is by far the most widely used. It is identified by a cross within a box and shows the area of oxide to be removed before the polysilicon is added. The area removed directly exposes the surface of the silicon so that the polysilicon directly connects to it.

Figure 5.6 shows examples of contact cuts. For a normal connection, polysilicon and n-diffusion connect over an area of 2λ×2λ and the contact cut extends at least one λ in all directions around the connection apart from the 2λ in the direction of the diffusion. For a depletion-mode transistor the minimum contact area is 4λ×4λ.

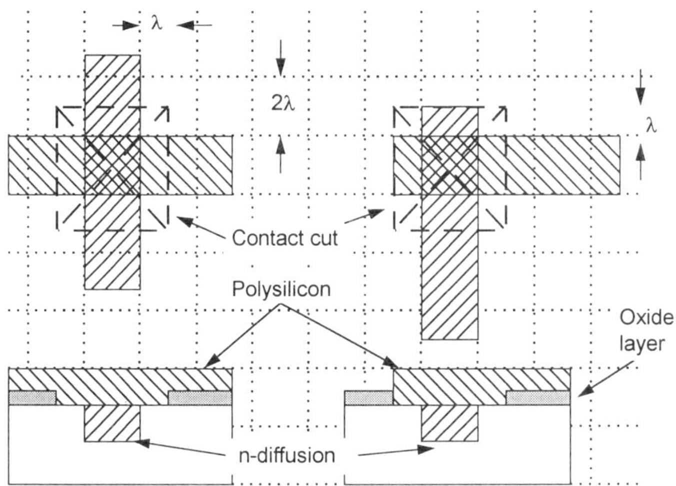

Figure 5.6 Contacts between polysilicon and n-diffusion.

5.6 Depletion-mode load transistor

A depletion-mode load transistor conducts when there is zero volts between the gate and source. To set up a zero voltage between the gate and source a buried contact connects to the diffusion layer and the polysilicon layer, as illustrated in Figure 5.7. The channel length then ranges from the upper part of the polysilicon to the upper part of the buried contact. In the transistor in Figure 5.7 the channel length is 4λ and its width is 2λ (L:W=2:1). The depletion region must extend at least 2λ above and below the polysilicon region.

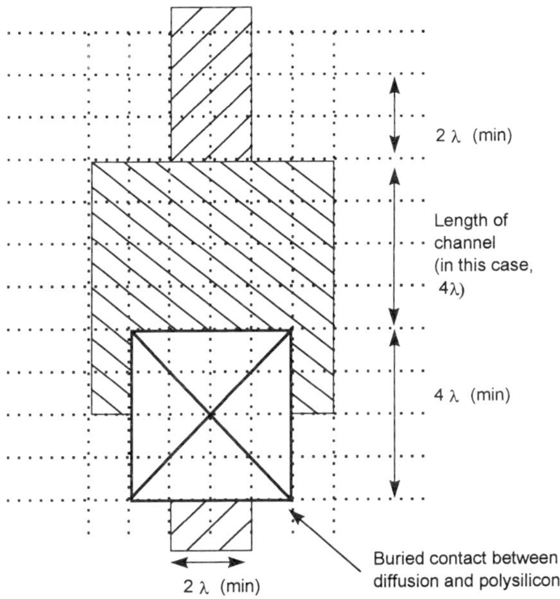

Figure 5.7 Layout of a depletion-mode transistor.

5.7 NMOS logic circuit design

The basic steps in drawing an NMOS stick circuit are as follows:

1. Draw metal for V_{DD} and GND.
2. Draw diffusion between rails and make contacts.
3. Connect depletion transistors with implant.
4. Connect enhanced mode transistors.

The following sections give some practical logic layouts.

5.7.1 Inverter

In order for an NMOS inverter circuit to provide a reliable low output state the ON resistance of the pull-up depletion-mode transistor must be at least four times greater than the resistance of the pull-down enhanced mode transistor. This can be set up by ensuring that the L:W ratio of the pull-up transistor is four times greater than the pull-down. Figure 5.8 shows a schematic and a stick diagram of this circuit.

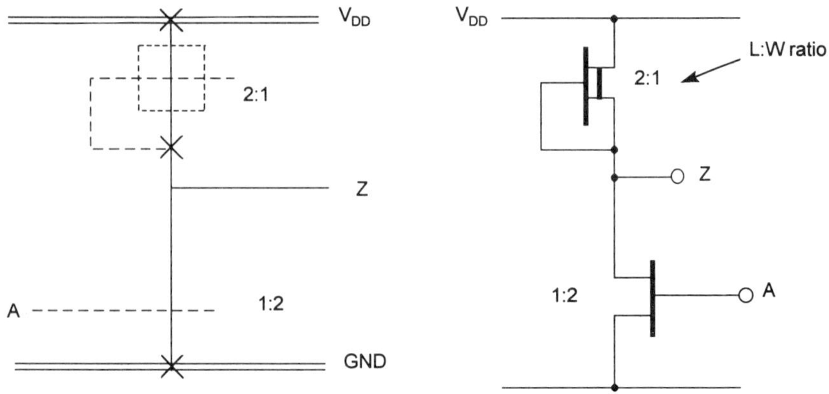

Figure 5.8 Inverter circuit and stick diagram.

Figure 5.9 shows how this gate could be laid-out using lambda design rules. The minimum width of the metal lines (V_{DD} and GND) are 3λ and the minimum width of polysilicon and diffusion lines are 2λ. The aspect ratio of the pull-down transistor is 1:2 (2λ:4λ) and the aspect ratio of the pull-up transistor is 2:1 (4λ:2λ). A via connects the diffusion layer to the metal layer.

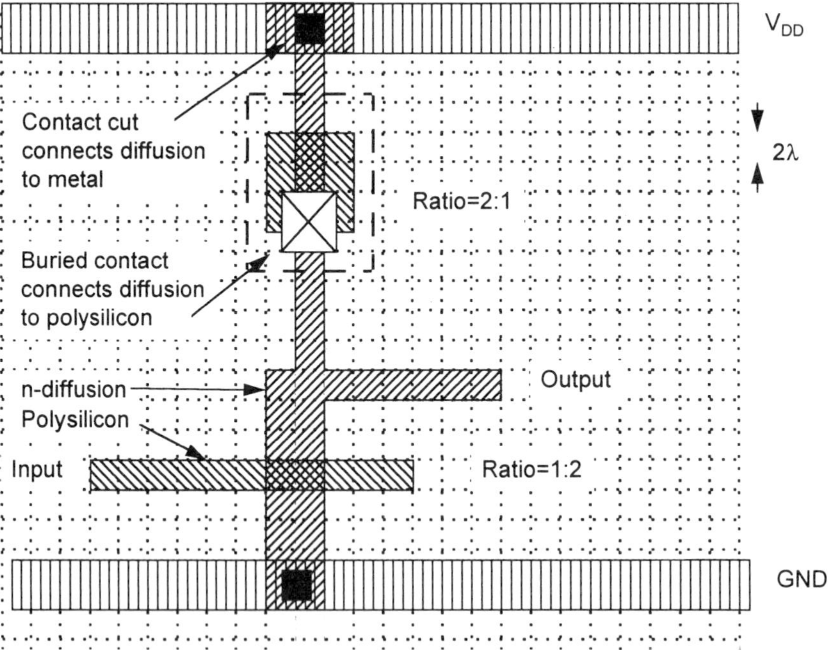

Figure 5.9 Layout of an NMOS inverter cell.

5.7.2 NOR gate

In order for an NMOS NOR circuit to provide a reliable low output state the ON resistance of the pull-up depletion-mode transistor must be at least four times greater than the resistance of the pull-down enhanced-mode transistor. This can be set up by ensuring that the L:W ratio of the pull-up transistor is at least four times greater than the pull-down. A schematic and a stick diagram of this circuit are given in Figure 5.10 and Figure 5.11 shows a possible silicon layout.

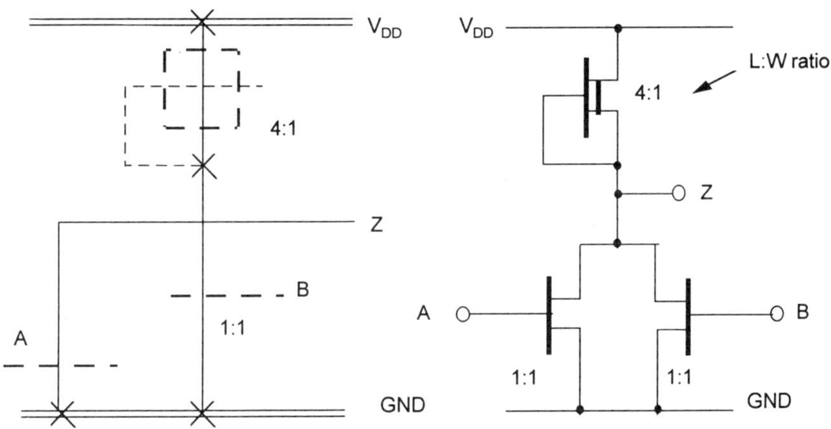

Figure 5.10 2-input NOR circuit and stick diagram.

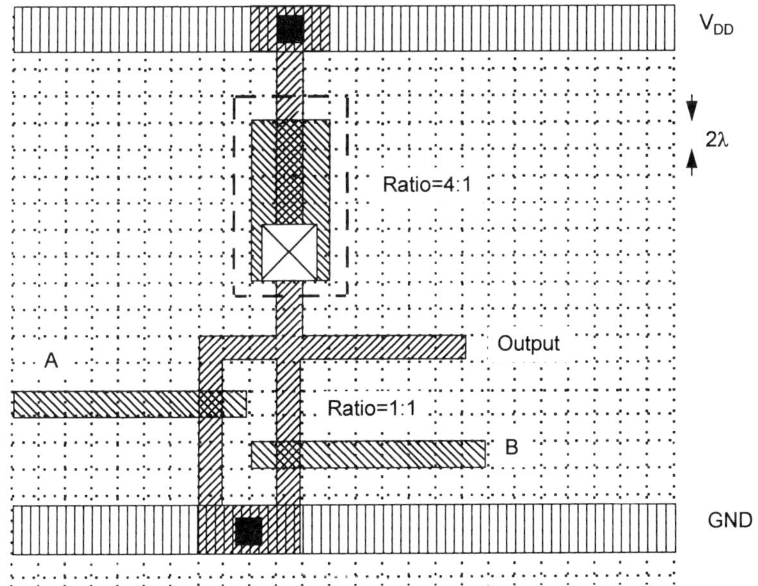

Figure 5.11 Layout of a 2-input NMOS NOR cell.

5.7.3 NAND gate

In order for a 2-input NMOS NAND circuit to provide a reliable low output state the ON resistance of the pull-up deletion-mode transistor must be at least four times greater than the combined resistance of the pull-up depletion mode transistors when they are ON. This can be set up by ensuring that the L:W ratio of the pull-up transistor is eight times greater than any of the pull-down transistors. Figure 5.12 gives the schematic and a stick diagram of this circuit and Figure 5.13 a possible silicon layout.

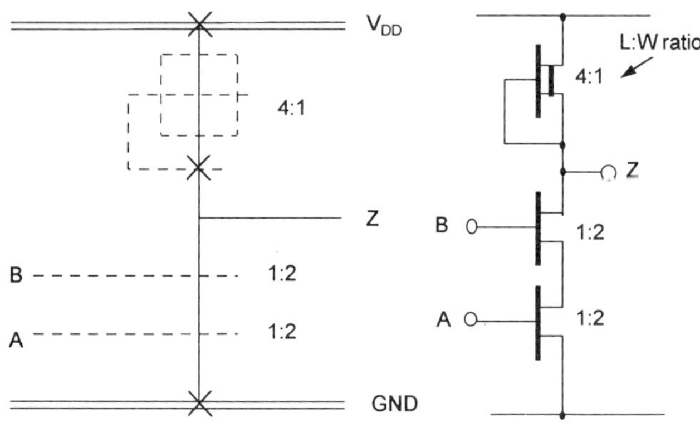

Figure 5.12 Layout of a MOS transistor.

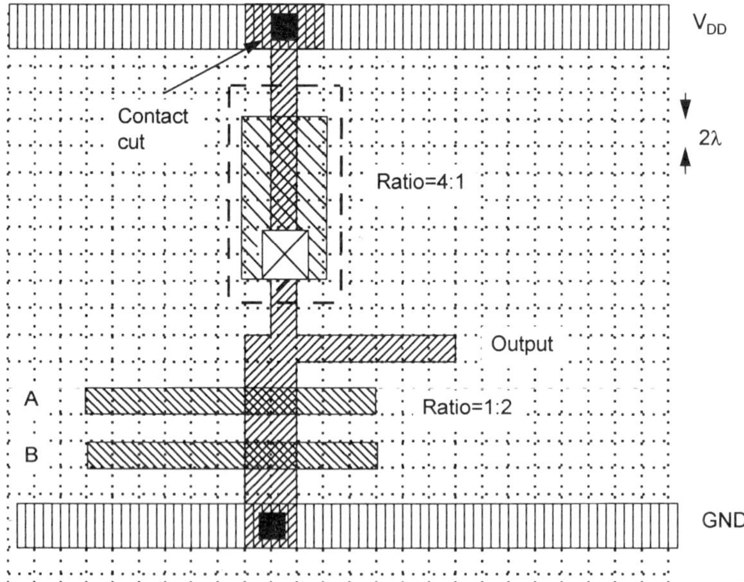

Figure 5.13 Layout of a 2-input NMOS NAND cell.

Up to this point all NMOS gates have been designed using NMOS lambda design rules and no account has been taken of actual resistance values. It has been stated that, in order to create a reliable output low level, for an inverter circuit, the pull-up resistance must be four times greater than the pull-down resistance. Many parameters depend on the actual resistance of the transistor channels, such as:

- the maximum drive current;
- gate timing characteristics, for example, the larger the channel resistance the slower the gate will respond;
- power supply drain; if the values of the channel resistance are too low then the drain on the power supply may be too great.

The channel of the NMOS transistor can be approximated to a rectangular block of conducting material having an area A, resistivity ρ, length W and thickness t, as shown in Figure 5.14.

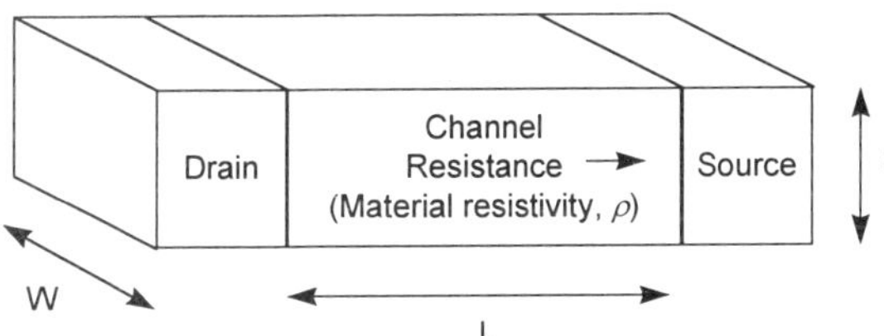

Figure 5.14 Approximation for an NMOS channel resistance.

The resistance will thus be approximated by:

$$R_{CH} = \frac{\rho L}{A} \quad \Omega$$

Since the cross-section area is $t \times W$, then

$$R_{CH} = \frac{\rho L}{tW} \quad \Omega$$

The resistance of the channel thus depends on the thickness of the channel and the resistivity of the diffusion. These two parameters are process depend, whereas, the length and the width of the channel are design dependent. Normally the channel resistance is related to a unit square of the material, that is, when W is equal to L, thus:

$$R_{CH} = \frac{\rho}{t} \quad \Omega \text{ per unit square}$$

The unit square resistance depends on the technology used. For 5 μm MOS transistor the value is around 10 000 Ω per unit square and for 1.2 μm technology it is around 20 000 Ω per unit square.

Note that since the units of length and width are the same their units cancel-out and the channel resistance is calculated by multiplying the ratio of the length to width by the unit square resistance. For example, the inverter in Figure 5.8 has a pull-up ratio of 2:1 and a pull-down of 1:2. Thus, for 5 μm MOS, the pull-up resistance will be approximately:

$$R_{UP} = R_{CH}\frac{L}{W} \quad \Omega$$
$$= 10^4.\frac{2}{1} \quad \Omega$$
$$= 20\ \text{k}\Omega$$

and the pull-down resistance will be approximately:

$$R_{DN} = R_{CH}\frac{L}{W} \quad \Omega$$
$$= 10^4.\frac{1}{2} \quad \Omega$$
$$= 5\ \text{k}\Omega$$

Thus the total resistance for a low output will be 25 kΩ. For a +5 V supply the drain current will be 0.2 mA (assuming no drain of current for the output drive). The equivalent circuit of this arrangement is given in Figure 5.15. The power dissipated in the circuit, for a low output, will thus be 1 mW ($5^2/25$ mW).

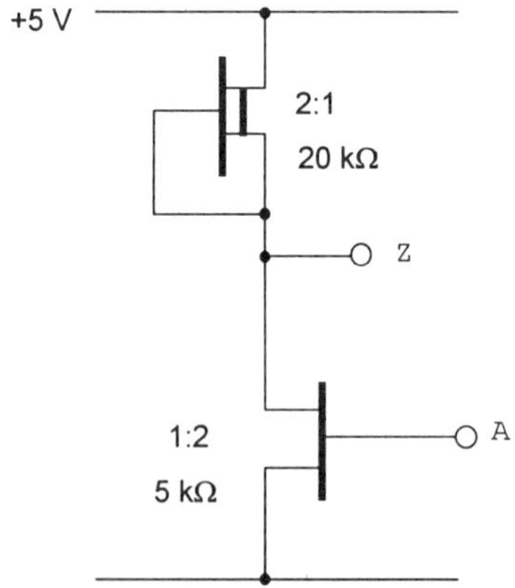

Figure 5.15 Approximate resistance values for inverter circuit.

5.9 Comparison of NAND and NOR gates

As has been seen the pull-down resistance of a NAND gate increases as the number of inputs. Thus, the greater the number of inputs to a NAND gate the slower the circuit becomes. This is because the pull-up resistance must be increased for each input to set up the 4:1 ratio of pull-up resistance to pull-down resistance.. For example, if the pull-down transistor is ratioed with 1:2, then a 2-input gate requires a 2:1 for the pull-up transistor, a 3-input gate requires an 4:1 for the pull-up, a 4-input gates requires an 8:1 for the pull-up transistor, and so on. This increasing resistance slows the gate because of RC time-constants.

NOR gates do not suffer from this problem as the pull-down resistance does not increase with the number of inputs.

5.10 Exercises

5.1 For the 2-input NAND and NOR gates given in the text, determine the equivalent resistances of the pull-up and pull-down transistors and thus determine the supply current when the output is low. Assume 5 μm MOS.

5.2 Design a stick diagram and silicon layout for the following gates:

(a) 3-input NOR gate;
(b) 3-input NAND gate;
(c) 3-input OR gate (3-input NOR gate connected to an inverter);
(d) 3-input AND gate (3-input NAND gate connected to an inverter).

5.3 For the circuits designed in Question 5.2 determine the equivalent resistances of the logic gates and determine the power supply drain current when the output level is low. Also determine the power dissipated in the logic gates for an output low.

5.4 Which of the following inverter gates dissipates the least power:

(a) Pull-up ratio of 4:1 and pull-down of 1:1, or
(b) Pull-up ratio of 2:1 and pull-down of 1:2.

CMOS Circuit Design

6

6.1 Introduction

CMOS circuits are used in many applications from gate arrays to control logic. They have the advantage over NMOS circuits in that they do not require active pull-up loads. For a CMOS inverter stage only one of the transistors conduct at a time. In the low output state the pull-down transistor conducts and the pull-up transistor is off. When the output is high the pull-up transistor conducts and the pull-down transistor is off. Pure CMOS circuits conduct very little current and, at low digital frequencies, consume small amounts of power. Most of the power dissipation is due to a changing logic state because both the pull-up and pull-down transistors are conducting. Thus, CMOS circuits dissipate an increasing power for increasing digital signal frequency.

6.2 CMOS processing

The three main types of CMOS processes are:

- *n-well CMOS* which uses a p-type substrate. As with NMOS technology, the NMOS transistors are formed by diffusing n-type material into the substrate. For the PMOS transistors, a well of n-type is diffused in the substrate and the p-type diffusion then defines the drain and source of the PMOS transistors. This type is compatible with NMOS technology as it uses a p-type substrate.
- *p-well CMOS* which uses an n-type substrate. PMOS transistors are formed by diffusing p-type material into the substrate. For NMOS transistors a well of n-type is diffusion followed by a p-type diffusion within the well. This technology is not as popular as n-well CMOS as it incompatible with NMOS fabrication.
- *Twin tub CMOS* which uses both n-type and p-type wells in a lightly doped substrate.

Figure 6.1 shows an n-well CMOS inverter layout. The substrate is p-type and the diffusion for the pull-down transistor is n-type (NMOS). To create a PMOS transistor a tub of n-type is diffused into the substrate. Next, p-type diffusion is inserted to create the drain and source of the transistor. The source of the pull-up transistor is joined to the drain of the pull-down by metal (as diffusion or polysilicon creates another transistor).

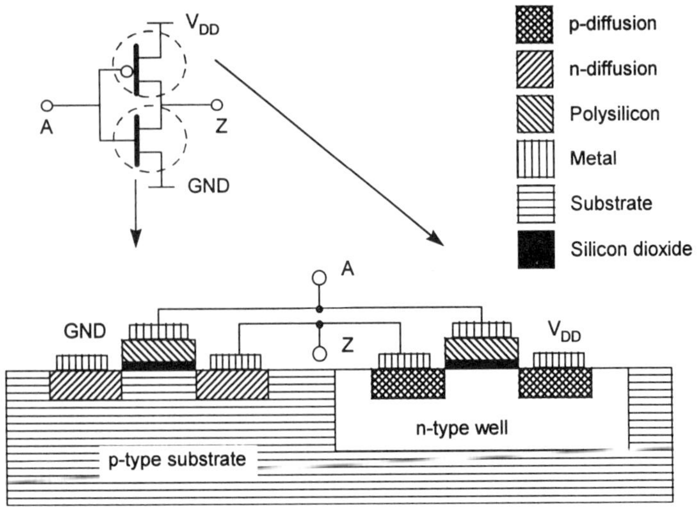

Figure 6.1 n-well CMOS inverter.

6.3 CMOS inverter

Figure 6.2 shows the stick diagram of a CMOS inverter gate. In this case, the input to the gate is polysilicon and the output connects to metal. The polysilicon crossing the p-type diffusion creates the PMOS pull-up transistor and polysilicon crossing the n-type diffusion creates the NMOS transistor. A dotted line shows the demarcation for the end of the p-well. Any diffusion in the p-well is p-type diffusion and below it is n-type. The current diffusion type is also described as the active diffusion and the connection between metal and the diffusion layer is named an active contact.

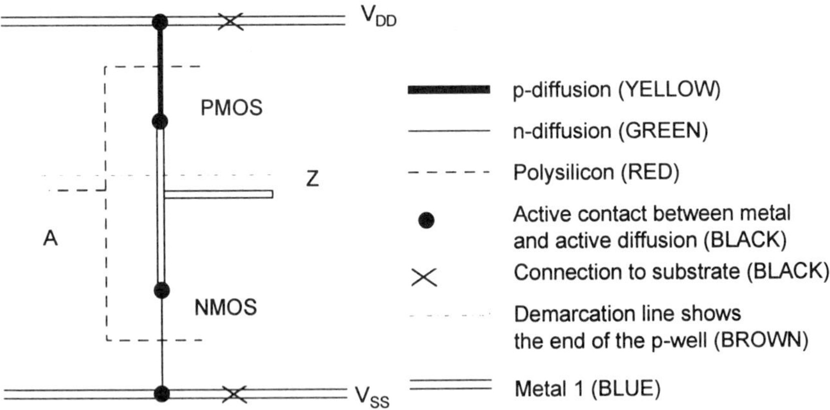

Figure 6.2 CMOS inverter.

As with NMOS, the stick diagram displays polysilicon as red, n-diffusion as green and the first metal layer as blue. For CMOS the p-diffusion layer is drawn in yellow and the demarcation region is shown in brown. Notice that there is a contact between the supply rails to the active diffusion (cross symbol) on each transistor. This is because the pull-up PMOS transistor requires the substrate to connect to V_{DD} and the pull-down NMOS transistor requires the substrate to be connected to V_{SS}.

Figure 6.3 shows a possible layout of this circuit. Notice that the minimum transistor size has been used (that is, $2\lambda \times 2\lambda$) and that the lengths and widths of the transistors do not have to be scaled as they were in NMOS design. This is because only one of the transistors is on in either logic state. Note that the connection between the substrate and the supply rails has not been included in Figure 6.2. This will be covered in the next chapter which will discuss practical CMOS cells.

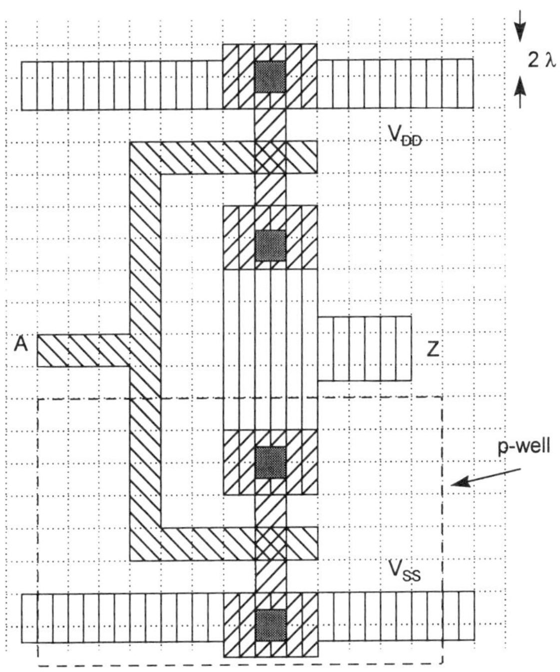

Figure 6.3 CMOS inverter.

6.4 CMOS NAND gate

A CMOS NAND gate requires two series pull-down NMOS transistors connected to ground and two parallel pull-up PMOS transistors connected to the supply voltage. Figure 6.4 gives a stick diagram. Only when there are two low inputs will the output go low (that is 00, 01, 10 gives a 1 and 11 gives a 0 output). Again, metal connects to the output and polysilicon as the input. If output metal layer from one gate connects to the next gate then the layer must be changed to polysilicon by inserting a contact between the metal and polysilicon layers. Figure 6.5 shows a possible layout of the circuit.

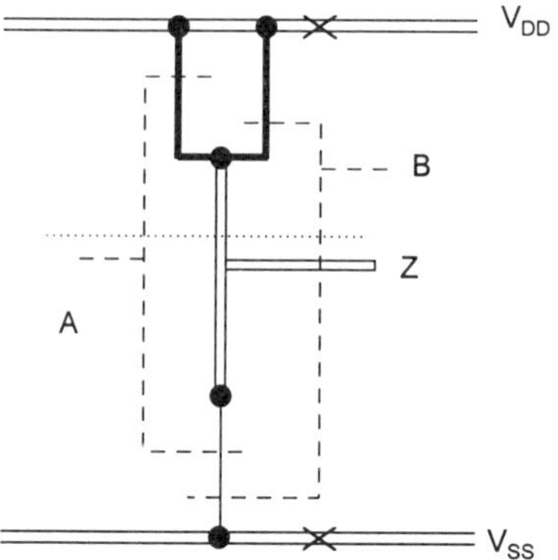

Figure 6.4 2-input CMOS NAND stick diagram.

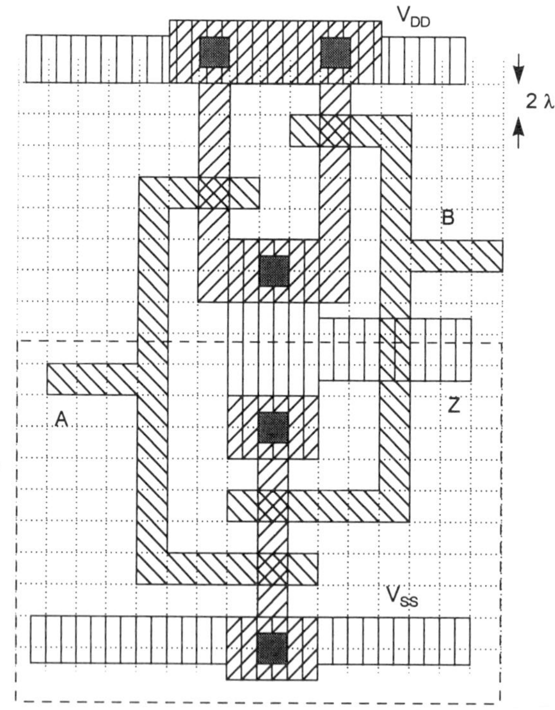

Figure 6.5 2-input CMOS NAND gate layout.

6.5 CMOS NOR gate

The CMOS NAND gate requires two parallel pull-down NMOS transistors and two series pull-up PMOS transistors. Figure 6.6 gives the stick diagram. Only when there are two lows on the inputs will the output go high (that is 00 gives a 1, and 01, 10 and 11 give a 0 output). Again, metal connects to the output and polysilicon to the input. Figure 6.7 shows a possible layout of the circuit.

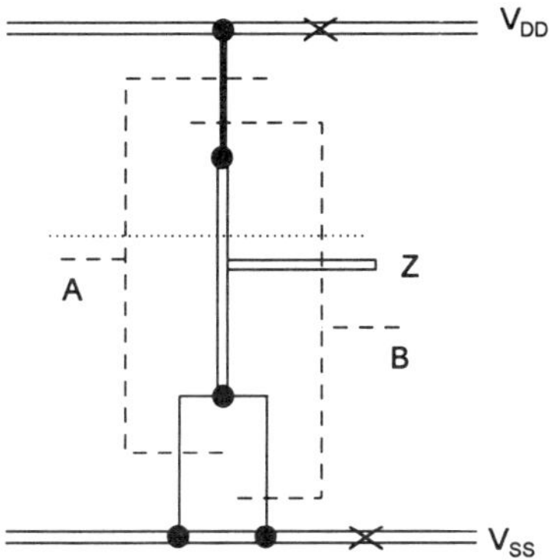

Figure 6.6 2-input CMOS NOR stick diagram.

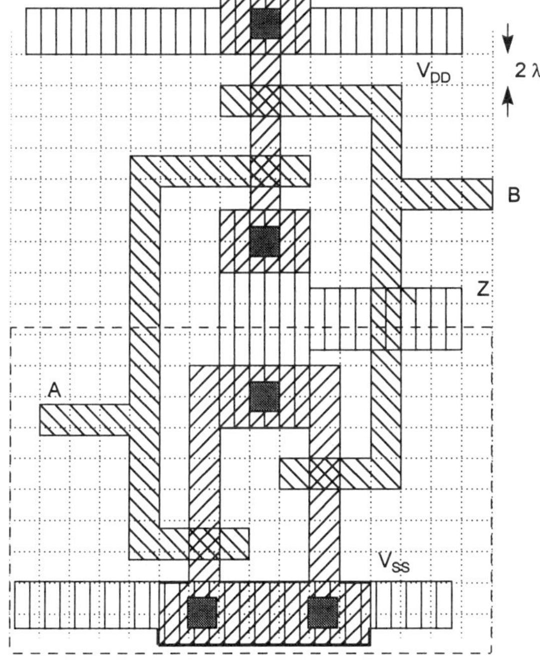

Figure 6.7 Layout of a 2-input CMOS NOR gate layout.

6.6 Electrical characteristics

The performance of CMOS circuits is not as dependent on the length to width ratio as only the pull-up or pull-down transistors conduct at a time. As was seen in the previous chapter the unit resistance of n-diffusion is approximately 10 kΩ per square for 5 μm technology and 20 kΩ per square for 1.2 μm technology. The p-type diffusion has a higher unit resistance and is approximately 25 kΩ per square (for 5 μm) and 40 kΩ per square (for 1.2 μm). Thus for an inverter, using 5 μm technology, with transistor ratios of 1:1 (2λ×2λ) the pull-up PMOS transistor will have a resistance of 25 kΩ and the pull-down resistance will be 10 kΩ. These resistances can be used to determine the maximum drive current in the low and high state, and also the power dissipation.

6.7 Exercises

6.1 Explain why the ratios of the pull-up and pull-down transistors with a CMOS inverter do not effect the output voltage levels. Contrast this with an NMOS inverter.

6.2 Explain what effect the L:W ratios of the CMOS inverter pull-up and pull-down transistors will have on circuit speed.

6.3 Determine the approximate cell sizes (in nλ×mλ) for the gates in Figure 6.3, Figure 6.5 and Figure 6.7.

6.4 Redesign the cells in Figure 6.3, Figure 6.5 and Figure 6.7 so that each of the cell size areas are minimized.

6.5 Using the approximations developed in Section 6.6 determine the short-circuit current of the following gates, for 5 μm technology with a supply of +15 V and 0 V:

(a) inverter;
(b) 2-input NAND gate;
(c) 2-input NOR gate.

Practical Design of an ASIC 7

7.1 Introduction

This chapter analyzes a practical IC and covers the top-level to bottom-level design of an ASIC (Application Specific Integrated Circuit). It uses the DIGCHIP design available with L-EDIT Version 5.00.

7.2 Requirements analysis

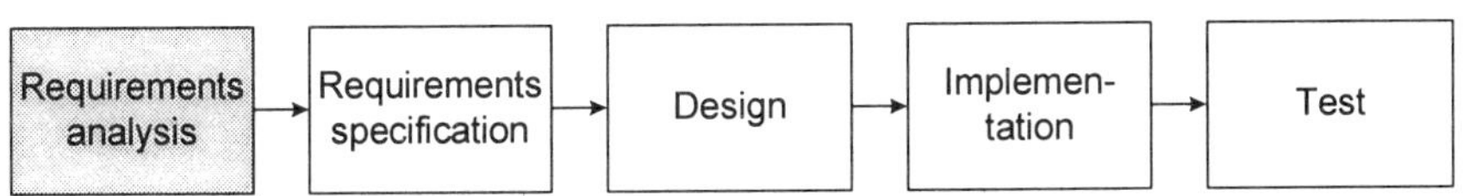

The requirements analysis and specification is the first stage of the process and is typically the most difficult part. Any failure in analyzing the requirements can be very costly in future work. At this stage, decisions are made as to the nature and bounds of the project and how the problem is to be solved. These decisions involve:

- stating how the problem is to be solved;
- stating the requirements that define successful solution of the problem.

If after this stage a requirement has been identified then a requirements specification is generated. This defines the technical requirements of the system and should state exactly how the system is intended to operate and all operational constraints.

In this case the objective is to produce an ASIC which will control a set of traffic lights and a WALK/DONT WALK light. A layout of the traffic light junction is shown in Figure 7.1. There are two sets of traffic lights, from North to South (_NS) and from East to West (_EW).

One complete sequence of a single traffic light is 64 seconds, made up of:

- 24 seconds GREEN.
- 8 seconds YELLOW.
- 32 seconds RED.
- *then repeat.*

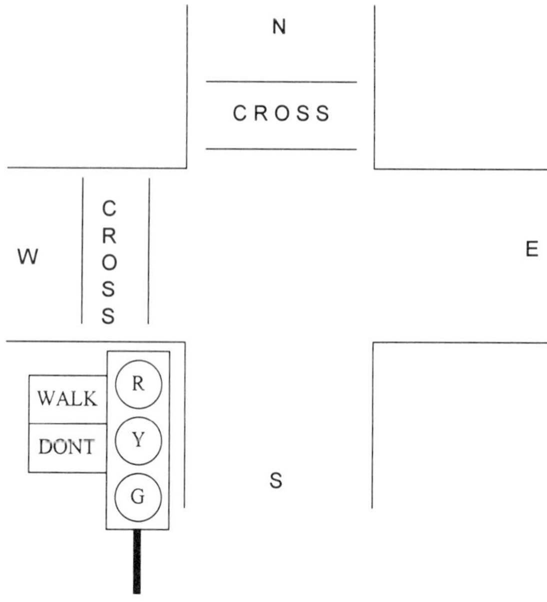

Figure 7.1 Traffic light junction.

With two traffic lights, one traffic light should show steady RED until the other traffic light has finished its sequence. Table 7.1 outlines the basic sequence.

When the East-West traffic light is showing RED then pedestrians can cross on the North-South junction and vice versa. Thus when the GREEN on the North-South junction is on then the North-South WALK light is on for 16 seconds. It will then flash for 8 seconds (16 flashes, that is on for 0.5 seconds and off for 0.5 seconds).

Table 7.1 Traffic light sequence.

Traffic light 1	*Traffic light 2*	*Time (seconds)*
GREEN	RED	24
YELLOW	RED	8
RED	GREEN	24
RED	YELLOW	8
GREEN	RED	24
YELLOW	RED	8
RED	GREEN	24
RED	YELLOW	8
and so on.		

7.3 Requirements specification

The requirement specification defines all the system requirements and any constraints or limitations. All functions performed by the system will be specified and all error and warning conditions are documented. The requirements specification is extremely important and all parties involved in the system development must agree on its content before a full design takes place.

7.3.1 Input/output requirements

The design requires the generation of a number of states which will be generated by a system clock. A reset input will also be required to start the system in a known state. These input signals will be named CLOCK and RESET. There are 12 outputs to the system for the traffic lights and walk/don't walk signals. These will be named RED_EW (for RED light on the East/West junction), YELLOW_EW, GREEN_EW, RED_NS, YELLOW_NS, GREEN_NS, WALK_EW, DONT_EW, WALK_NS and DONT_NS. Figure 7.2 shows the input/output requirements and for the ASIC.

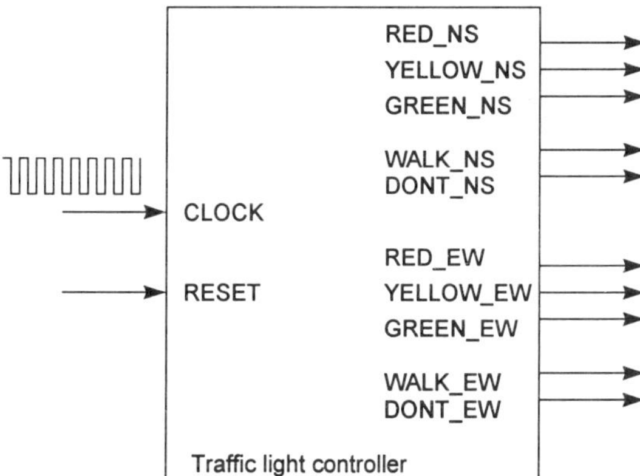

Figure 7.2 Connections to system.

7.3.2 Waveform diagram

It can be shown that the system requires a total of 128 states. This will require a 7-bit counter. Figures 7.3 to 7.5 show the waveform diagram for the system. After this the waveform repeats. It can be seen that since that the system clock must operate at 8 cycles over 4 seconds, or 2 Hz.

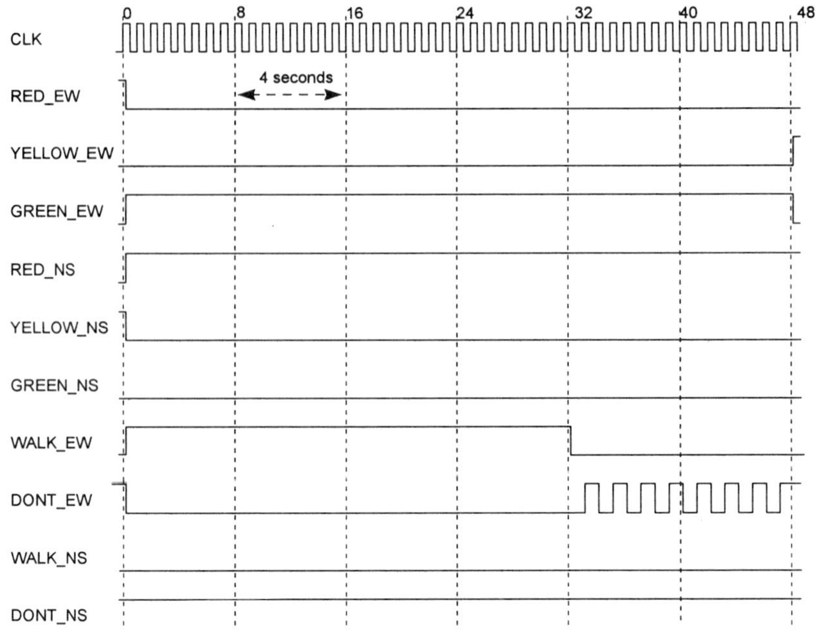

Figure 7.3 Waveform from 0 to 48 clock pulses.

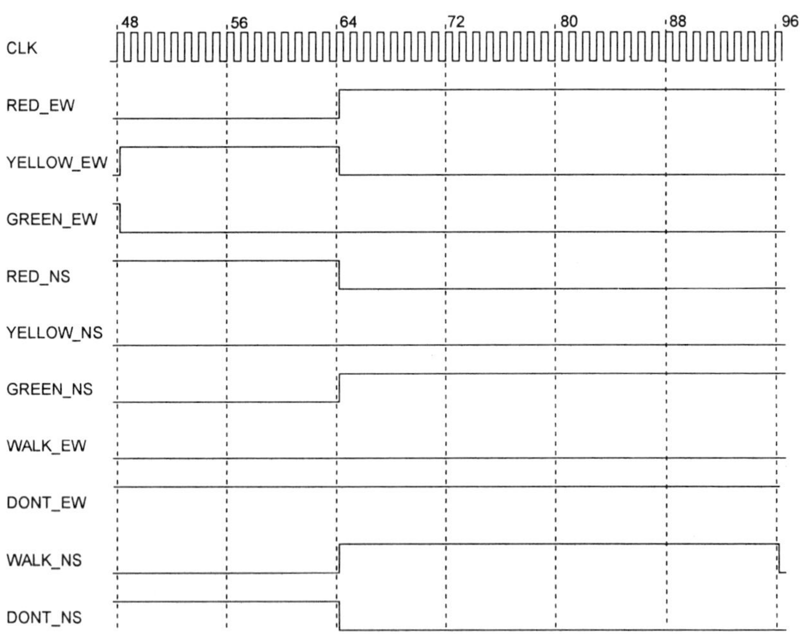

Figure 7.4 Waveform from 48 to 96 clock pulses.

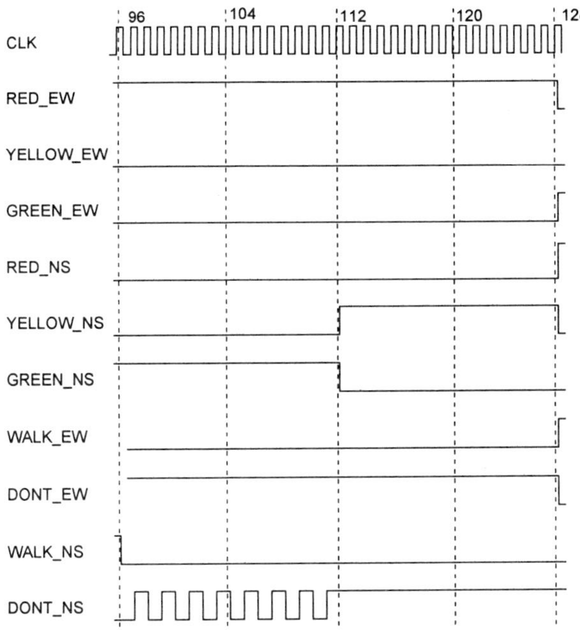

Figure 7.5 Waveform from 96 to 128 clock pulses.

7.3.3 State table

From the waveform diagrams in Figures 7.3 to 7.5 a state table can be produced. An outline of one for this design is given in Table 7.2 (the completion of the complete table is left as an exercise). The period between each change of state is 0.5 seconds. The signals have been assigned shortened names, such as R_E (RED_EW) and D_E (for DONT_EW).

Table 7.2 Traffic light sequence.

State	*R_E*	*Y_E*	*G_E*	*R_N*	*Y_N*	*G_N*	*W_E*	*D_E*	*W_N*	*D_S*
0000000	0	0	1	1	0	0	1	0	0	1
0000001	0	0	1	1	0	0	1	0	0	1
:::	:::	:::	:::	:::	:::	:::	:::	:::	:::	:::
0100000	0	0	1	1	0	0	0	0	0	1
0100001	0	0	1	1	0	0	0	1	0	1
0100010	0	0	1	1	0	0	0	0	0	1
0100011	0	0	1	1	0	0	0	1	0	1
0100100	0	0	1	1	0	0	0	0	0	1
0100101	0	0	1	1	0	0	0	1	0	1
0100110	0	0	1	1	0	0	0	0	0	1
0100111	0	0	1	1	0	0	0	1	0	1
:::	:::	:::	:::	:::	:::	:::	:::	:::	:::	:::
1111110	1	0	0	0	1	0	0	1	0	1
1111111	1	0	0	0	1	0	0	1	0	1

7.4 Design

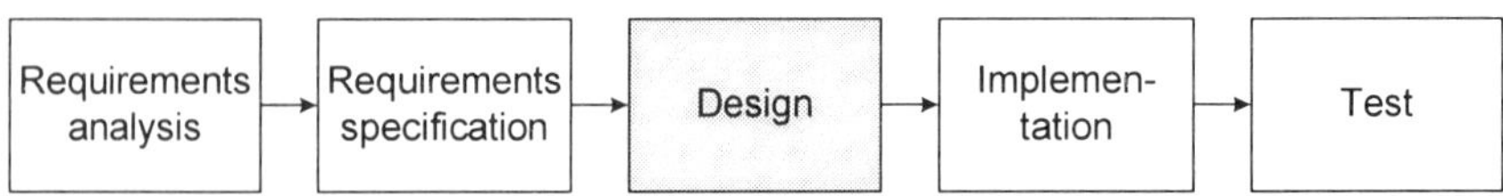

Figure 7.6 shows a top-level representation of the traffic light controller. The 7-bit counter produces an output 0000000 to 1111111 on the signal lines GFEDCBA. This output is then fed into combinational logic to produce the required logic levels. The K-map for the GREEN_EW signal is given in Table 7.3.

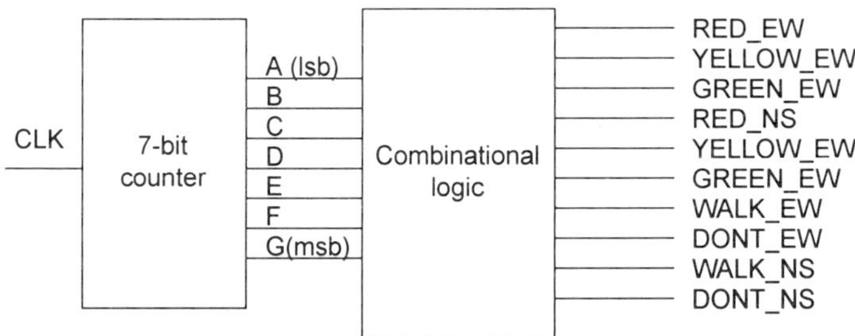

Figure 7.6 Top-level representation of the traffic light controller.

Table 7.3 K-map for GREEN_EW signal.

	CBA							
GEFD	000	001	011	010	110	100	101	111
0000	1	1	1	1	1	1	1	1
0001	1	1	1	1	1	1	1	1
0011	1	1	1	1	1	1	1	1
0010	1	1	1	1	1	1	1	1
0110	0	0	0	0	0	0	0	0
0100	0	0	0	0	0	0	0	0
0100	1	1	1	1	1	1	1	1
0111	1	1	1	1	1	1	1	1
1111	0	0	0	0	0	0	0	0
1110	0	0	0	0	0	0	0	0
1101	0	0	0	0	0	0	0	0
1001	0	0	0	0	0	0	0	0
1000	0	0	0	0	0	0	0	0
1011	0	0	0	0	0	0	0	0
1010	0	0	0	0	0	0	0	0
1100	0	0	0	0	0	0	0	0

From the K-map it can be seen that the GREEN_EW signals will give:

$$GREEN\_EW = \overline{G}.\overline{F} + \overline{G}F.\overline{E}$$

This can then be minimized to produce:

$$GREEN\_EW = \overline{G}.(\overline{F} + F.\overline{E})$$

which can be further minimized to give:

$$GREEN\_EW = \overline{G}.(\overline{F} + \overline{E})$$

The conversion of the state table to the resultant logic is left as an exercise. The resulting logic is:

$$RED\_EW = G$$
$$RED\_NS = \overline{G}$$
$$YELLOW\_EW = \overline{G}.F.E$$
$$YELLOW\_NS = G.F.E$$
$$GREEN\_EW = \overline{G}.\overline{F} + \overline{G}.\overline{E}$$
$$GREEN\_NS = G.\overline{F} + G.\overline{E}$$
$$WALK\_EW = \overline{F}.\overline{G}$$
$$DONT\_EW = G + A.F.\overline{E} + \overline{G}.F.E$$
$$WALK\_NS = G.\overline{F}$$
$$DONT\_NS = \overline{G} + A.F.\overline{E} + G.F.E$$

From these logic equations the resultant electronic design can be generated. A 7-bit ripple counter will be used. Figure 7.7 shows the resultant schematic after some minimization of terms.

7.5 Simulation

All designs must be simulated to prove that they are error-free. In this case it is assumed that the design is fault-free and the actual simulation of the circuit is left as a tutorial example. Appendix D contains a C program which simulates the sequence of operations. An ECAD package such as OrCAD, Cadence or Mentor Graphics can be used to simulate the hardware operation. The results should be check against the requirements specification.

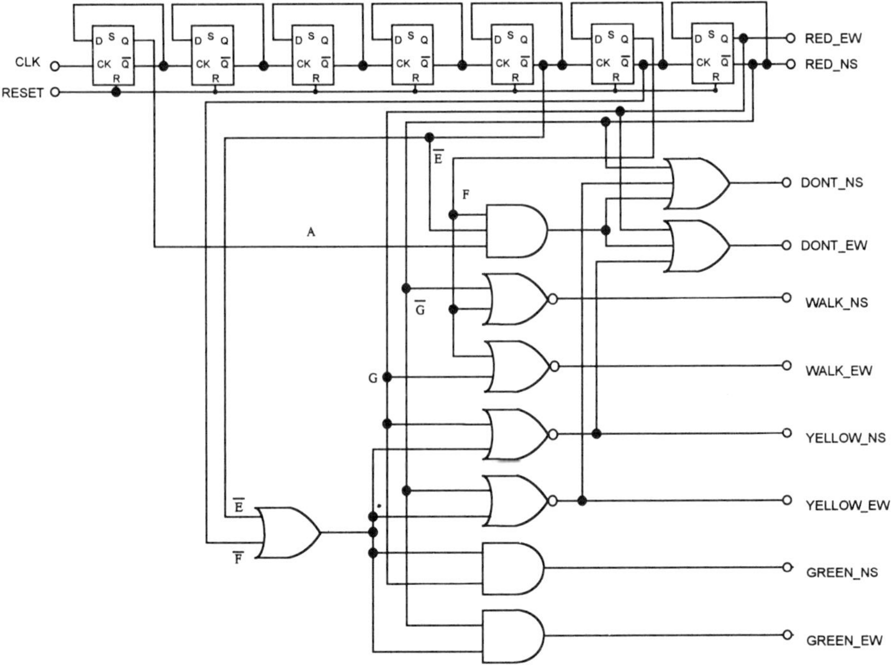

Figure 7.7 Traffic light controller schematic.

7.6 Implementation

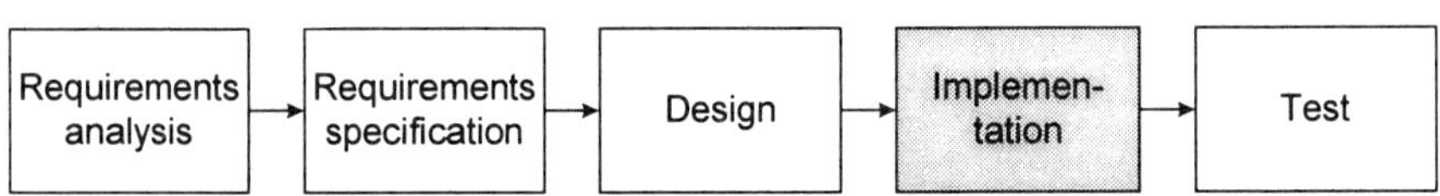

This section analyzes the DIGCHIP ASIC. Figure 7.8 shows the connection to the core of the ASIC. These connections are bounded to the plastic package with gold wires. Other signals are also used that are not included in the schematic in Figure 7.7. These are TOP4 (D output on 7-bit counter), TOP5 (E output on 7-bit counter), TOP6 (D output on 7-bit counter), and TST_PNT ($\overline{E}$ output on 7-bit counter), There are a total of 18 pins which would probably connect to a 18-pin DIL package.

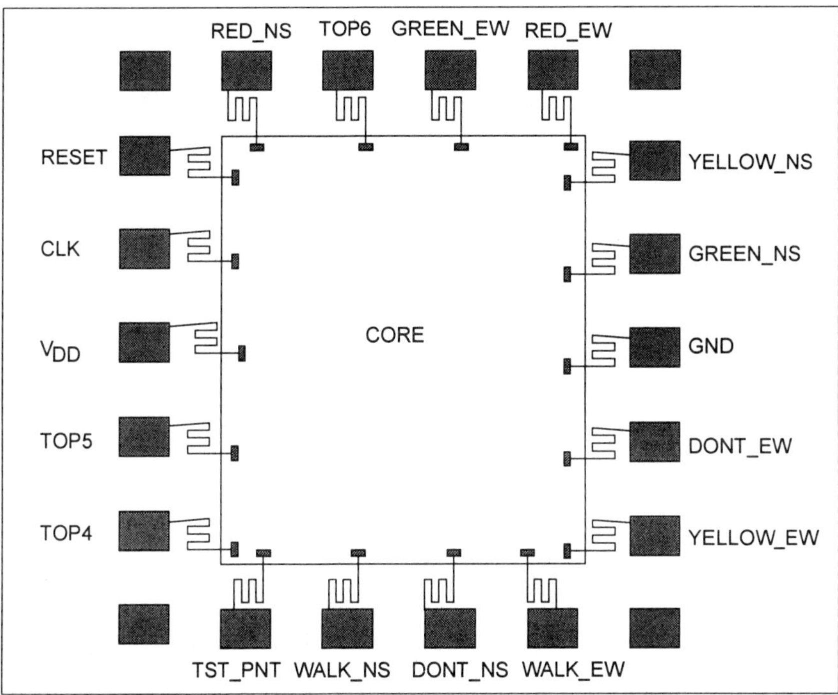

Figure 7.8 Traffic light controller schematic.

7.6.1 Top-level

Figure 7.9 shows the completed ASIC. The design uses two layers of metal. From the input pads the V_{DD} and GND lines connections connect with one layer of metal and the other input/output signals connect from the core to the pads with the second layer of metal. Figure 7.10 shows how the power rails connect to the logic cells. The top of the cells connects to V_{DD} and the bottom of the cells to GND. It can also be seen from Figure 7.10 that the other signal lines cross underneath the power supply lines.

It can be seen from Figure 7.9 and Figure 7.11 that the logic cells are arranged in three rows. The signal connection from the pads to the rows are shown around the outside of the diagram. Connections between the logic cells are made in the channels between the rows (the interconnection highway). These are made with metal connections, horizontal with metal 1 (so as not to short-circuit to the supply rails) and vertical for metal 2. The metal 1 layer is shaded darker than the metal 2. A via (represented as a black box) connects layer 1 to layer 2. The interconnection of the cells is illustrated in Figure 7.12.

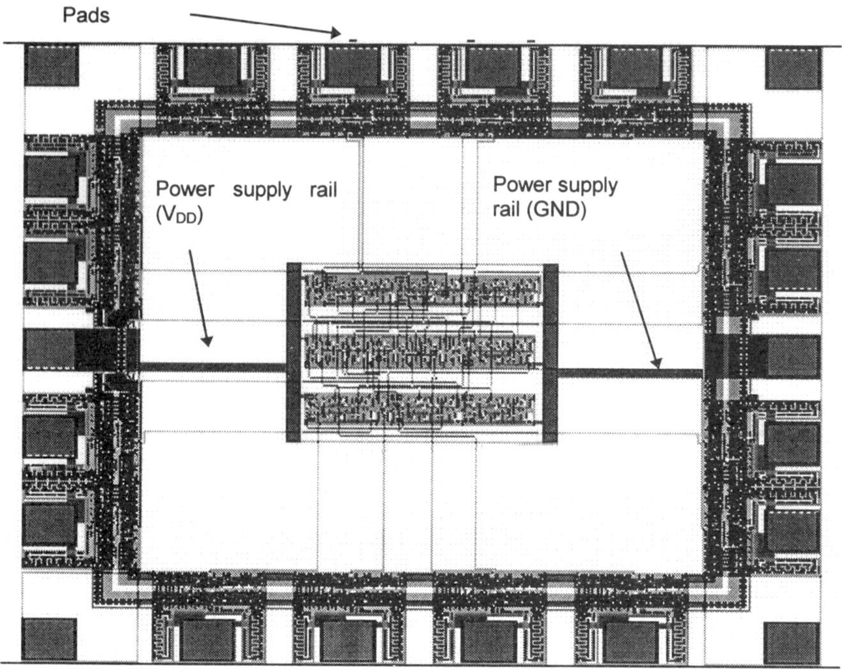

Figure 7.9 Traffic light ASIC.

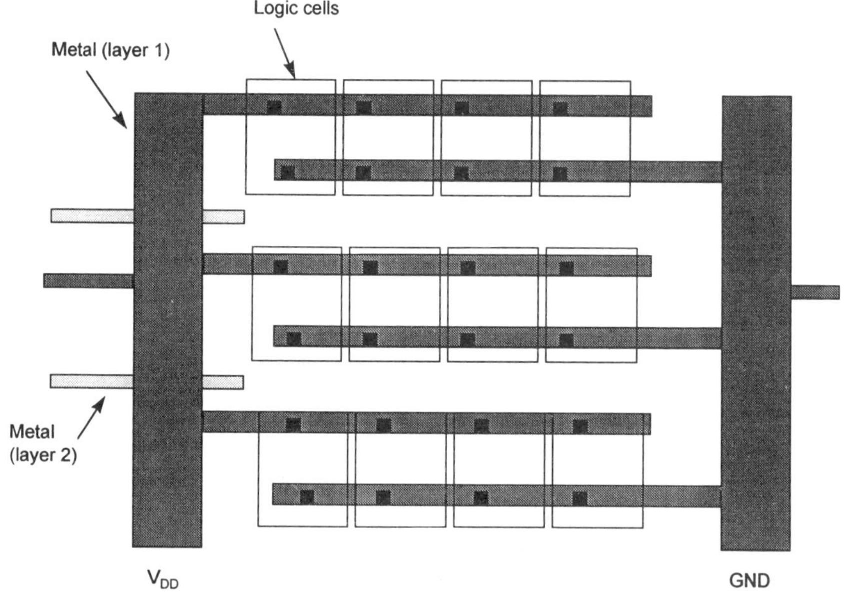

Figure 7.10 Power supply connections.

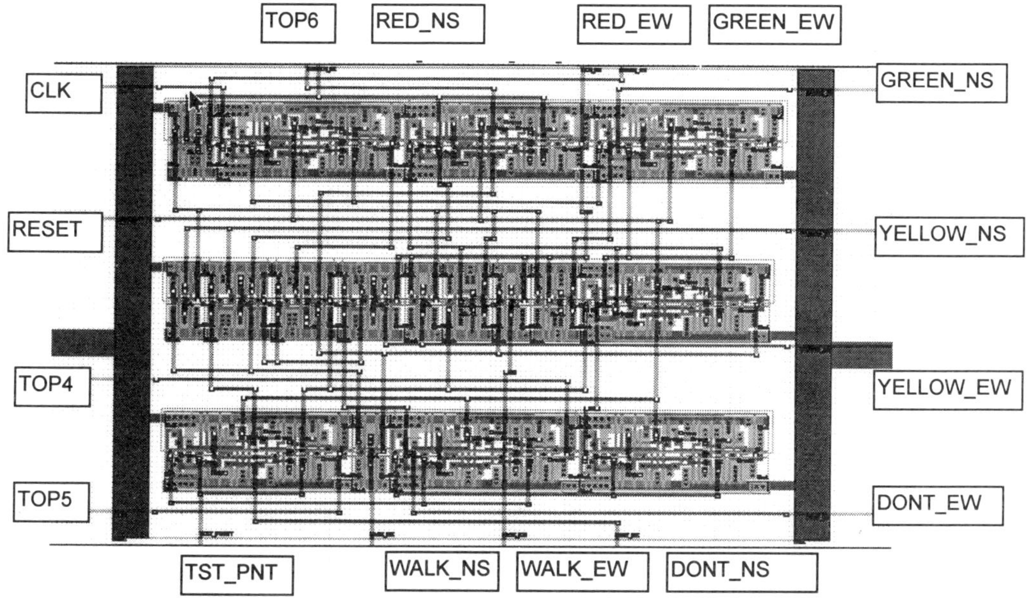

Figure 7.11 Traffic light ASIC showing the three rows of cells.

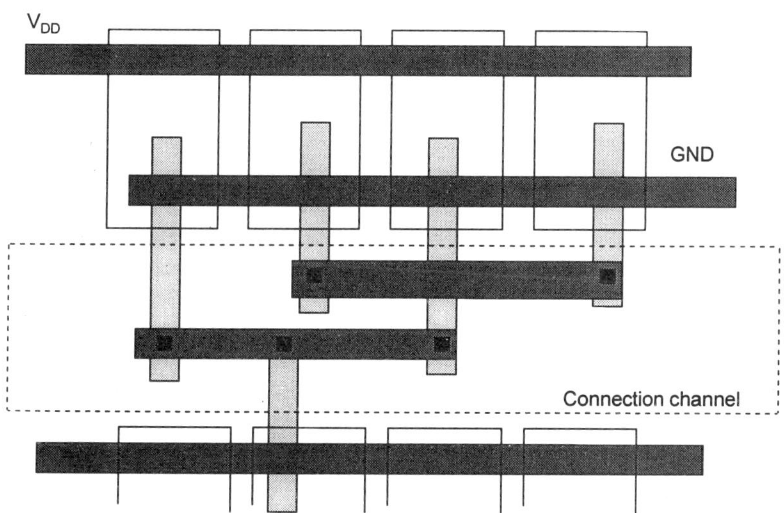

Figure 7.12 Connections between cells with metal 1 and metal 2.

The arrangement of the logic cells is given in Figure 7.13. A 2AND represents a 2-input AND gates, DFF represent a D-type flip-flop, 2NOR represents a 2-input NOR gate, and so on.

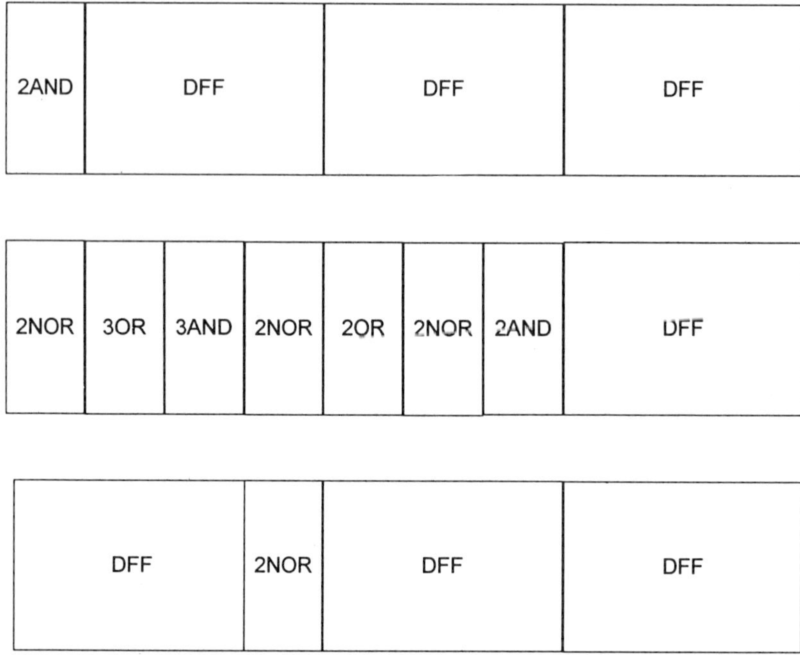

Figure 7.13 Layout of logic cells.

7.6.2 Logic cell design

This section discusses the design of the 2-input NOR gate. The analysis of the remaining cells will be left as an exercise. Figure 7.14 shows the design 2-input NOR gate. It differs in its shape from the cell designed in the previous chapter because it has been compacted to save space. The equivalent transistor circuit on the left-hand side shows that the gate equates to a 2-input NOR gate. Inputs to the circuit are A, B and the output is Out. These connect to the cell with the metal 2 layer. The white boxes show the connection from metal 1 to metal 2 and the black boxes show the connection between metal and diffusion (an active contact) or between metal and polysilicon (a polysilicon contact). The width of the polysilicon is 2λ, the total length of the cell is 70λ and its width is 26λ.

Space is saved by connecting the metal of the supplies into the diffusion layer (rather than vice versa, in the previous chapter). The polysilicon line runs vertical through the cell and crosses the diffusion layer to create the required transistor layout.

The best way to understand the circuit is to trace the current flow from VDD to GND. First the current travels along the metal 1 layer and makes contact to the first PMOS transistor on the upper left-hand side of cell (the PMOS transistor connected to A). The current then flows through the diffusion horizontally and encounters the first polysilicon crossing (which makes the PMOS transistor). If this line is a low level then the current will cross the transistor and encounter another polysilicon crossing (the PMOS transistor connected to the B input). Again if the B input is a low then it will cross the channel in the diffusion on the other side. Metal 1 then connects to the diffusion and the current then flows downward and to the left-hand side. Next it connects to metal 2 by a via. The metal 2 layer directs the current downwards where it connects to metal 1 again which then connects to the diffusion layer in the lower part of the cell. The current can then take two routes (if the A and B input are high). Metal connects the diffusion back to the GND rail.

Figure 7.15 gives the equivalent stick diagram (which is easier than cell layout to trace the logical flow).

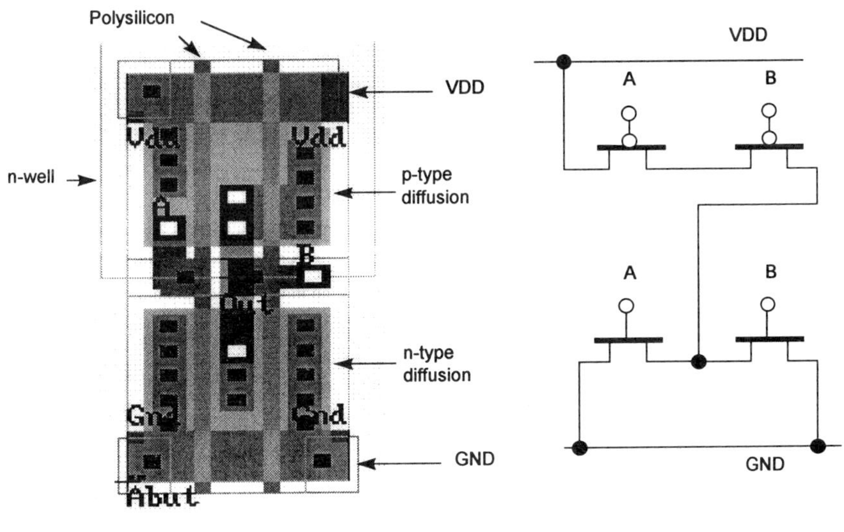

Figure 7.14 2-input NOR gate cell and equivalent transistor circuit.

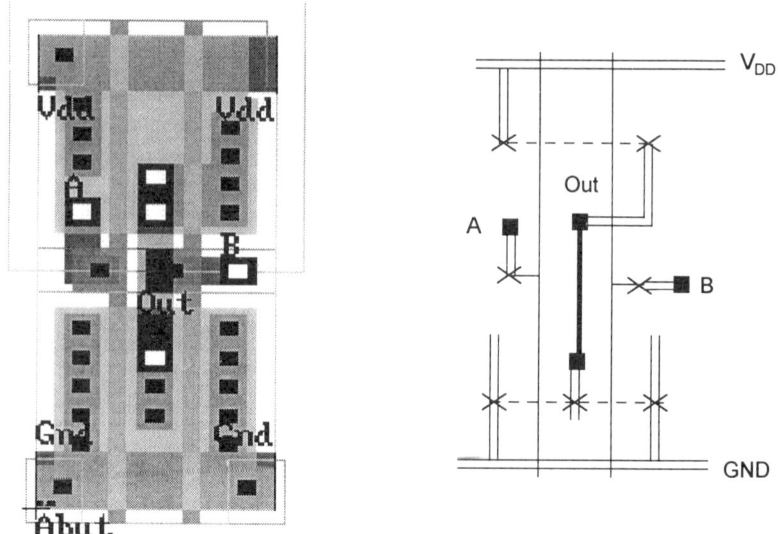

Figure 7.15 2-input NOR gate cell and equivalent stick diagram.

Figure 7.16 shows the layout and dimensions of the two layers of metal. Figure 7.17 shows the masks for diffusion and polysilicon layers. All vias and contacts are 2λ×2λ.

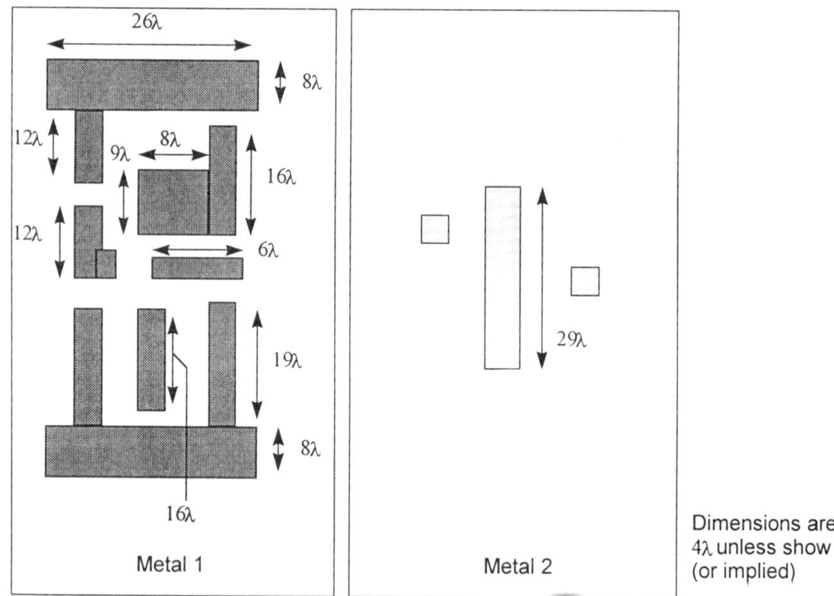

Figure 7.16 Metal 1 and metal 2 masks for 2-input NOR gate.

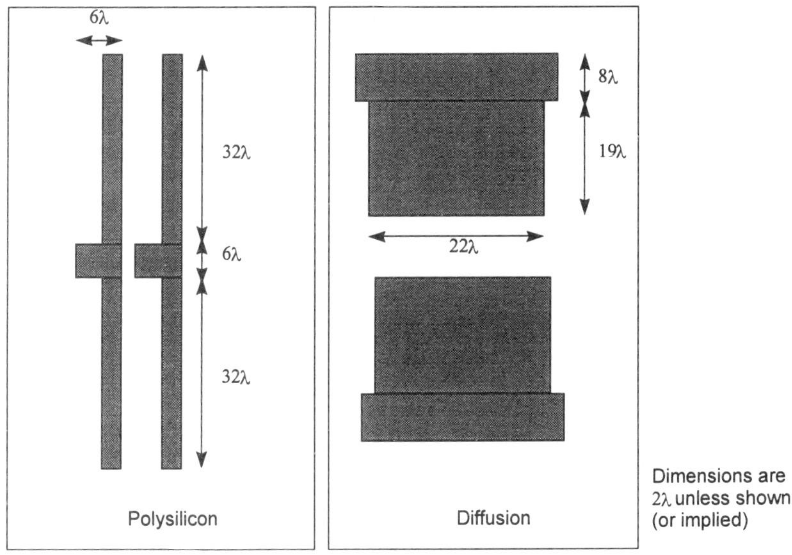

Figure 7.17 Polysilicon and diffusion masks for 2-input NOR gate.

Figure 7.18 shows a cross-section of the cell taken in lower-half of the cell where there is a via and two contacts between metal 1 and the diffusion.

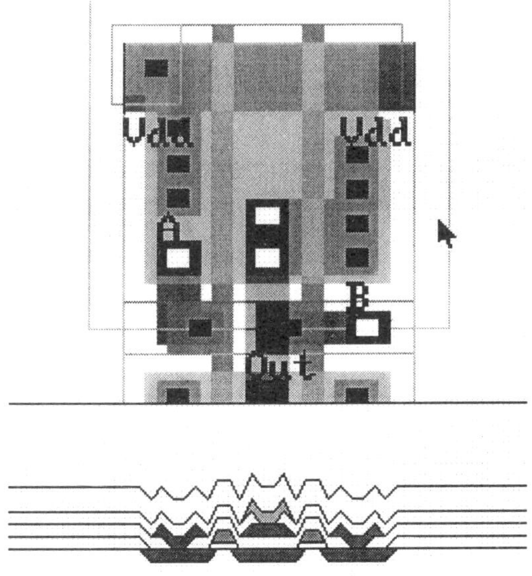

Figure 7.18 Cross section of 2-input NOR gate showing contacts between layers.

7.7 Exercises

7.1 Explain how the power supply lines connect to individual logic cells and also explain how the signal lines connect from the pads to the core of the cell.

7.2 Explain how the logic cells electrically interconnect.

7.3 Complete the state table in Table 7.2 for all of the states.

7.4 The complete set of Boolean equations for the traffic light design is given next. Derive each of them.

$$RED\_EW = G$$

$$RED\_NS = \overline{G}$$

$$YELLOW\_EW = \overline{G}.F.E$$

$$YELLOW\_NS = G.F.E$$

$$GREEN\_EW = \overline{G}.\overline{F} + \overline{G}.\overline{E}$$

$$GREEN\_NS = G.\overline{F} + G.\overline{E}$$

$$WALK\_EW = \overline{F}.\overline{G}$$

$$DONT\_EW = G + A.F.\overline{E} + \overline{G}.F.E$$

$$WALK\_NS = G.\overline{F}$$

$$DONT\_NS = \overline{G} + A.F.\overline{E} + G.F.E$$

7.5 With reference to Figures 7.11 and 7.13 identify the location of each of the gates in the schematic in Figure 7.7.

7.6 Figure 7.19 is a 2-input NOR gate with an inverted output. Out1 is the NOR output and Out2 is the OR output. Identify its operation and each of the layers. If possible, estimate the lengths and widths of each of the layers.

7.7 Figure 7.20 is a 2-input NAND gate with an inverted output. Out1 is the NAND output and Out2 is the AND output. Identify its operation and each of the layers. If possible, estimate the lengths and widths of each of the layers.

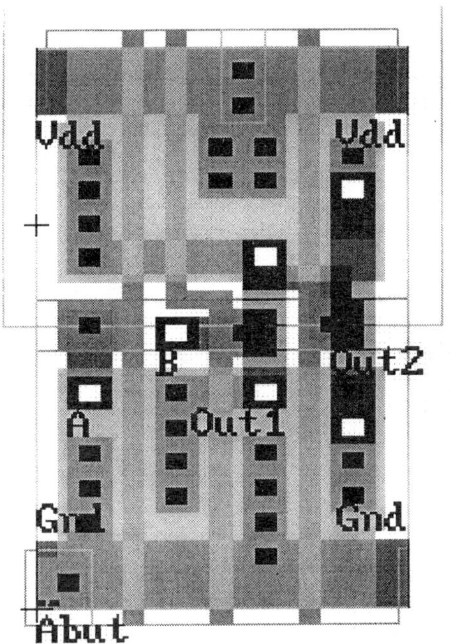

Figure 7.19 2-input NOR with inverter output.

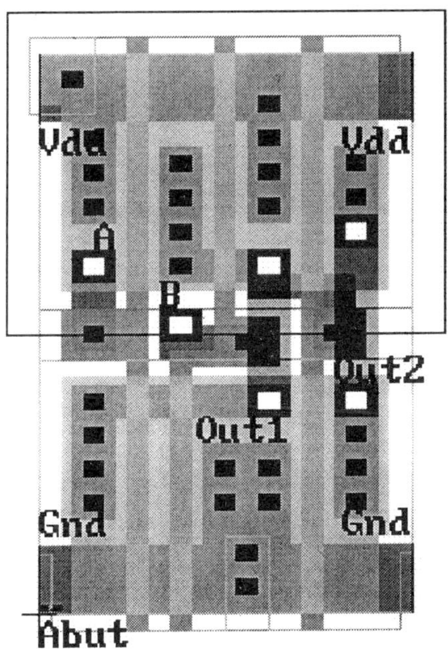

Figure 7.20 2-input NAND with inverter output.

7.8 Figure 7.21 is a 3-input NOR gate with an inverted output. Out1 is the NOR output and Out2 is the OR output. Identify its operation and each of the layers. If possible, estimate the lengths and widths of each of the layers.

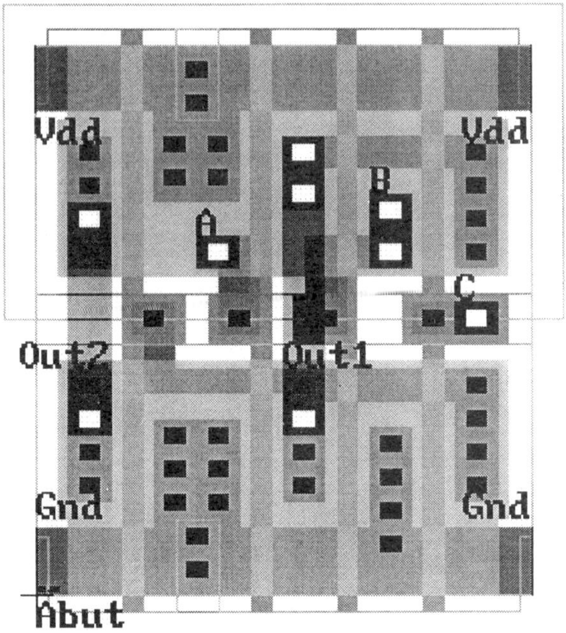

Figure 7.21 3-input NOR with inverted output.

7.9 Investigate the D-type gate and draw its equivalent transistor schematic (if you have access to the design package).

7.10 Investigate other designs for the traffic light controller, such as having a separate gated oscillator for the flashing and reducing the number of states on the counter.

7.8 Projects

7.8.1 Project 1: Traffic light controller 1

Design an ASIC for a road junction with a single traffic light. The sequence and timing of each of the individual lights is outlined in Table 7.4.

A WALK light will be active after the traffic light has completed a full sequence. This should stay active for 4 seconds, followed by a flashing WALK light for another 4 seconds (on for 1 second and off for 1 second). When this goes OFF a DON'T WALK light should be ON.

Table 7.4 Traffic light timings.

Sequence No.	*Sequence*	*Timing (seconds)*
1	RED	6
2	RED/YELLOW	1
3	GREEN	4
4	YELLOW	2
1	RED	6
2	RED/YELLOW	1
	and so on.	

7.8.2 Project 2: Traffic light controller 2

Design an ASIC for a road junction with two traffic lights. The sequence and timing of lights is outlined in Table 7.5.

A WALK light will be active after the traffic light has completed a full sequence. This light activates when pedestrians are allowed to cross the road and should stay active for 4 seconds, followed by a flashing WALK light for another 4 seconds (on for 1 second and off for 1 second). When this goes OFF the DON'T WALK light should be ON. Note that the WALK/DONT WALK lights are common to both traffic lights.

Table 7.5 Traffic light timings.

Sequence No.	*Sequence (traffic light 1)*	*Sequence (traffic light 2)*	*Timing (seconds)*
1	RED	RED	6
2	RED	RED/YELLOW	1
3	RED	GREEN	4
4	RED	YELLOW	2
5	RED	RED	6
6	RED/YELLOW	RED	1
7	GREEN	RED	4
8	YELLOW	RED	2
1	RED	RED	6
2	RED	RED/YELLOW	1
	and so on.	*and so on.*	

Memory Systems

8

8.1 Introduction

Binary information can either be stored, and recalled, electronically in memory circuits, or by using a mechanical device, such as an optical or hard disk. Mechanical storage methods are normally much slower than electronic methods. For example, an optical disk can transfer data at a rate of around 300 KB/second whereas an electronic memory can transfer data at more than 100 MB/second. Mechanical storage devices are normally used for mass storage of permanent data. An optical disk can store over 600 MB and a hard disk over 1 GB.

Electronic memory circuits consist of arrays of transistors. The three main types of memory are:

- Random access Read-Only Memory (ROM);
- Static Random Access Memory (SRAM);
- Dynamic Random Access Memory (DRAM).

A ROM stores data permanently and does not lose its contents when the power is taken away, whereas RAM is a volatile memory. The two main types of RAM are SRAM and DRAM. SRAM is normally faster than DRAM but it uses at least six transistors to store just one bit of information. DRAM, on the other hand, only requires a single transistor to store a single data bit. Thus, for a given amount of transistors, DRAM memory stores at least six times the amount of data than an equivalent SRAM memory. Unfortunately, DRAM suffers two main drawbacks: they require to be refreshed with power several hundred times per second and are slower than equivalent SRAM memories.

Memory circuits have address, data and control lines, as illustrated in Figure 8.1. The number of address lines indicates how many addressable locations there are and the number of input/output data lines shows how many data bits can be accessed at a time. For example, a 1 MB memory has 20 address lines, 2 MB has 21 address lines and a 4 MB memory has 22 address lines.

When the Select line goes active (that is, LOW) the memory IC is activated. The Read line loads data from the memory, whereas, the Write writes data into the memory. The address location of the memory to be read from, or written to, is defined by the address on the address lines. This address must be placed on the address bus before activating either the Read or Write lines.

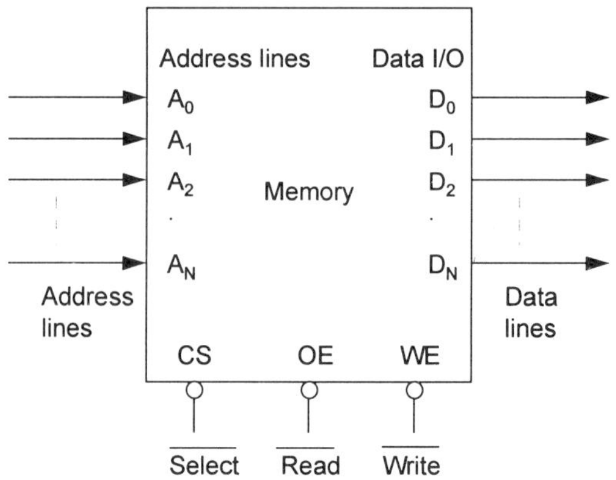

Figure 8.1 RAM memory connections

8.2 Memory addressing size

The size of the address bus indicates the maximum addressable number of bytes. Table 8.1 shows the size of addressable memory for a given address bus size. For example:

- A 1-bit address bus can address up to two locations (that is 0 and 1).
- A 2-bit address bus can address 2^2 or 4 locations (that is 00, 01, 10 and 11).
- A 20-bit address bus can address up to 2^{20} addresses (1 MB).
- A 24-bit address bus can address up to 16 MB.
- A 32-bit address bus can address up to 4 GB.

Table 8.1 Addressable memory (in bytes) related to address bus size.

Address bus size	*Addressable memory (bytes)*	*Address bus size*	*Addressable memory (bytes)*
1	2	15	32K
2	4	16	64K
3	8	17	128K
4	16	18	256K
5	32	19	512K
6	64	20	1M†
7	128	21	2M
8	256	22	4M
9	512	23	8M
10	1K*	24	16M
11	2K	25	32M
12	4K	26	64M
13	8K	32	4G‡
14	16K	64	16GG

* 1K represents 1024
† 1M represents 1 048 576 (1024 K)
‡ 1G represents 1 073 741 824 (1024 M)

8.3 ROM

ROM is a permanent memory and is typically used to store programs and system data. A PC uses a ROM to store the BIOS (Basic I/O System) software. BIOS is a set of basic I/O procedures that allows the system to use the keyboard, monitor, hard disk, and so on. The BIOS also contains the software required to start the system (the bootstrap program).

There are four main types of ROM: mask-programmed ROM, erasable programmable ROM (EPROM), programmable ROM (PROM) and EEPROM (electrically erasable programmable ROM). A mask-programmed ROM has a fixed structure and cannot be erased or reprogrammed. It is normally only cost-effective when producing large quantities.

An EPROM is programmed by an EPROM programmer. It can be reprogrammed by first erasing the data by exposing it to a high-intensity ultraviolet light (the EPROM device has a transparent quartz window). A PROM contains fusable links to store data. It is programmed by a PROM programmer but, unlike the EPROM, it cannot be erased. The best solution for flexibility is an EEPROM (or E^2PROM) which is programmed and erased electrically, typically without removing it from the installed system. Most modern PCs use E^2PROMs for their BIOS as they have the advantage of being updated by software. The main types of ROM are thus:

- ROM - permanent memory which cannot be erased. Only economical in large quantities.
- PROM - programmed by a PROM programmer, but cannot be reprogrammed.
- EPROM - programmed using an EPROM programmer and can be unprogrammed using UV light.
- EEPROM - programmed and erased electrically.

Some typical EPROMs are:

- 27256 (256Kb or 262 144 bits);
- 27512 (512Kb or 524288 bits);
- 27101 (1024Kb or 1 048 576 bits).

An n-bit addressable ROM has up to 2^n address locations, each of which stores a number of data bits. For example, a 9-bit addressable ROM has 512 different addressable locations, as illustrated in Figure 8.2. A decoder converts the 9-bit address code into one of 512 different address locations. In this case the ROM stores 8 bits for each address location. The presence, or absence, of a transistor defines a stored bit pattern. These transistors are either fixed (as in a ROM), are sensitive to UV light (as in an EPROM) or can be set up electrically (as in an E^2PROM). When a 0 is stored then the line connecting the output to the address line is connected to ground via a transistor. If a 1 is stored there is no pull-down transistor present to ground the output line. Figure 8.3 shows a 9-bit addressable ROM with a 5-bit output. The pull-up transistors give a HIGH when there is no pull-down transistor present. In this case memory location 0 stores the bit pattern `11111`, location 1 stores `01110`, location 2 stores `10111`, and so on to location 511 which stores `10111`.

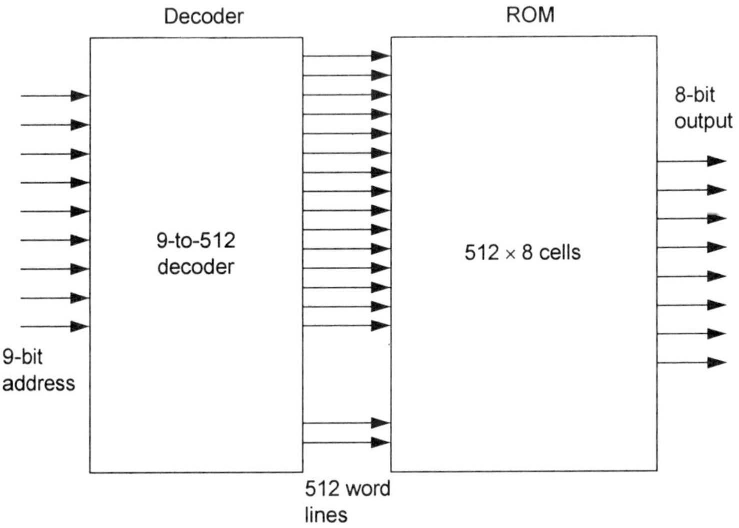

Figure 8.2 ROM addressing.

8.4 2D addressing

As the number of address lines increases the larger the address decoder becomes. This is unmanageable for large memories. For example, a 1 MB memory would require over a million addressable memory locations (requiring a 20-to-1 048 576 decoder). An improved memory arrangement is to use a matrix of rows and columns. This technique is described as 2D addressing and is used to reduce the address decoder size.

Figure 8.4 shows an 11-bit addressable memory. Address lines A0-A6 accesses 128 rows of data. Each of these rows contains 128 bits of data. The address lines A7-A10 then select 1 of 16 lines to give the data lines D0-D7 (128 bits divided by 16 addressable lines gives 8 data bits). It contains 128×128 memory cells giving a total of 65 386 memory cells arranged into 16×128 locations in memory (2 048 locations). Each location holds 8 bits of data and the total storage is thus 2 KB.

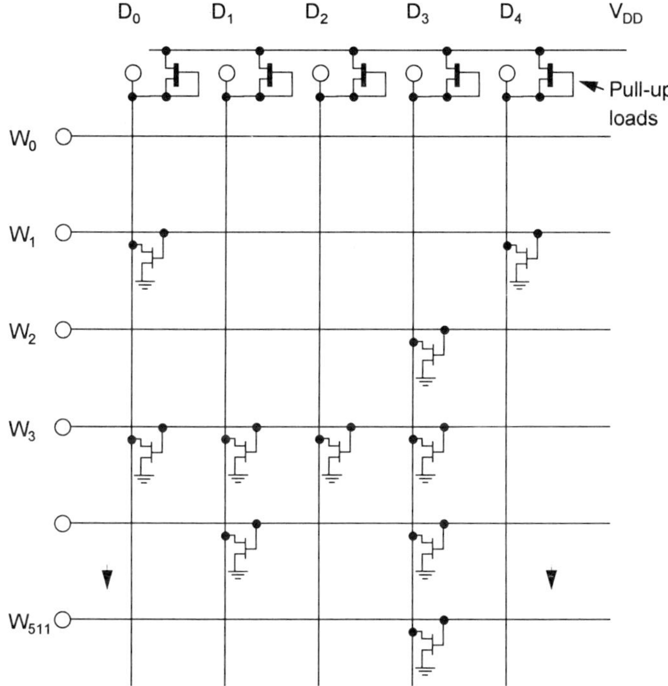

Figure 8.3 9-bit addressable NMOS ROM array.

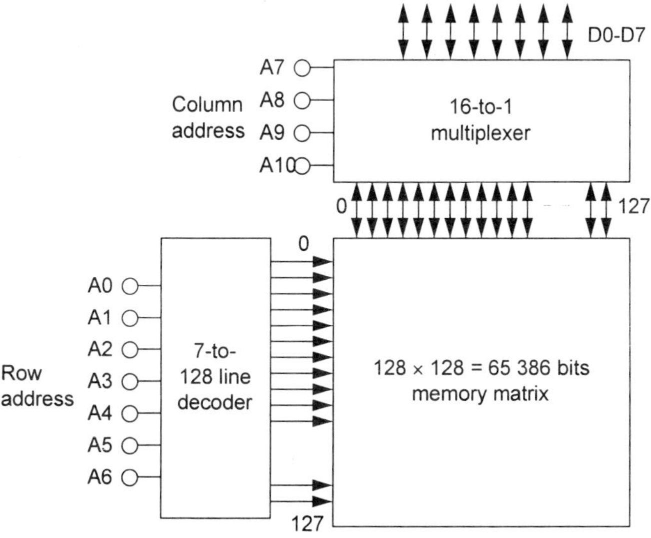

Figure 8.4 2D addressing of a 16 Kbit memory (2 KB or 2 048 × 8 bits).

8.5 Static RAM (SRAM) memory

Static RAM (SRAM) memory retains its data for as long as the power is applied. This type of memory uses a number of transistors arranged as flip-flop storage cells. Each stored bit requires at least six transistors and it is thus difficult to get as much memory as a DRAM IC. Typically the maximum available memory on a single IC is 128 kB and is normally found in memories that are less than 512KB

Figure 8.5 shows a typical static RAM cell. Transistors Q_3 and Q_4 are pull-up transistors and are either NMOS depletion-mode or CMOS p-type transistors (depending on whether the technology is NMOS or CMOS). To store a bit the cell is selected with the Cell Select line (that is, it is set active HIGH). If bit is HIGH then Q_2 will be ON and Q_1 OFF as its gate is held LOW by Q_2. A LOW on the Bit line causes Q_1 to be OFF and Q_2 ON. When the Cell Select line is placed inactive the cell stores the state.

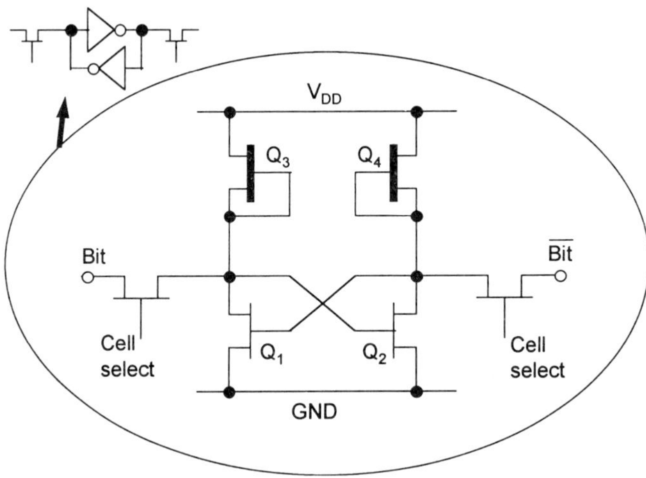

Figure 8.5 Static NMOS RAM cell.

8.6 Dynamic RAM (DRAM) memory

A DRAM cell uses fewer transistors than an equivalent SRAM cell and it is thus possible to get a greater amount of DRAM memory on a single IC. This is because DRAM uses small capacitors to store electrical charge, a typical cell is shown in Figure 8.6. Unfortunately, the capacitor loses its charge over a period of time and it must thus be refreshed every 2 to 4 ms. Normally the DRAM array is configured so that a refresh automatically occurs internally for any read or write operation. Thus, because of the continual refreshing and the loss of charge a DRAM IC consumes more power than a comparable SRAM. Another disadvantage is that the capacitors take a finite time to charge and discharge. This causes DRAMs to be considerably slower than SRAMs.

Memory arrays are made using DRAM cells arranged in rows and columns. A 256 × 256 (64 Kbit) array is shown in Figure 8.7. This requires 256 rows and 256 columns.

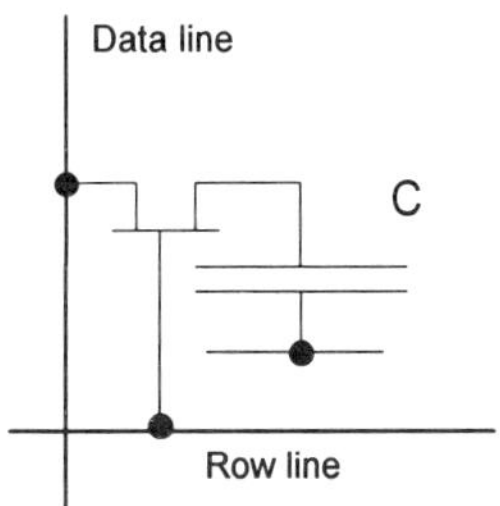

Figure 8.6 Basic Dynamic RAM cell.

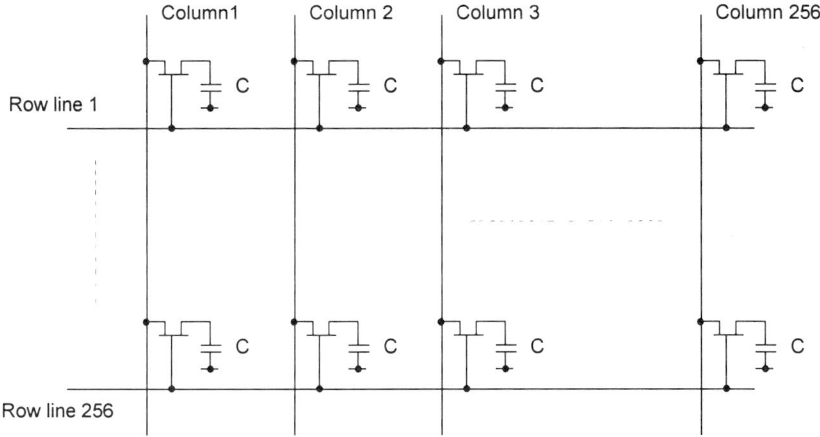

Figure 8.7 Basic DRAM memory array (256 × 256 bits = 65 536 bits = 64 Kbit).

A DRAM cell is automatically refreshed every time it is read from or written to. It would be difficult to read the contents of every memory cell so DRAM cells are constructed in a way that a single read from memory refreshes a whole row of bits at the same time. The DRAM array in Figure 8.8 has 1 024 cells in each row. When a single row is selected all 1 024 cells within it are automatically refreshed. Thus, in this case, only 256 refreshes are required to update the whole memory. These reads typically occur every 4 ms – one from each internal row.

The array in Figure 8.8 is a 256 Kb × 1 DRAM memory. The address lines A0-A7 select one of 1 024 rows. These lines are fed into four 1-of-256 multiplexers. The address lines A8-A15 then select one bit from the each of the 1-to-256 multiplexers. One bit is then selected using the address bits A16 and A17. If the single bit is being written to the array then the R/W line is a HIGH and the input on Din is read. A LOW on the R/W line writes the bit to the output line Dout.

To reduce the number of pins on the device the address lines are reduced so that the row and column are latched in two separate operations. The RAS line allows the row address to be latched and CAS the column address. This is shown in Figure 8.9. A 256Kb × 1 DRAM array thus only requires 16 pins (A0-A8, RAS, CAS, DIN, DOUT, R/W, VCC and GND).

DRAMs can be mounted on a single plug-in unit called a SIMM (single in-line memory module). A 30-pin SIMM is shown in Figure 8.10. It has a 10-bit address line input (A0-A9), an 8-bit data output/ input (D0-D7).

An example SIMM is the TMS024 which has 8 × 1 Mbit DRAM ICs, each of which provide one bit of the output (or input). It stores 1 MB of data and requires 512 refresh cycles over a period of 8 ms. A higher specification memory module is the 72-pin SIMM; Appendix H lists its pin connections.

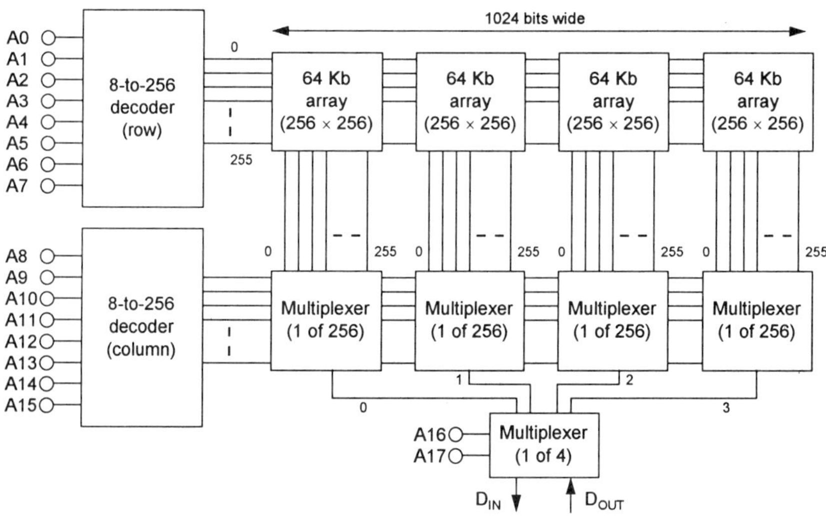

Figure 8.8 256 Kbit × 1 DRAM array.

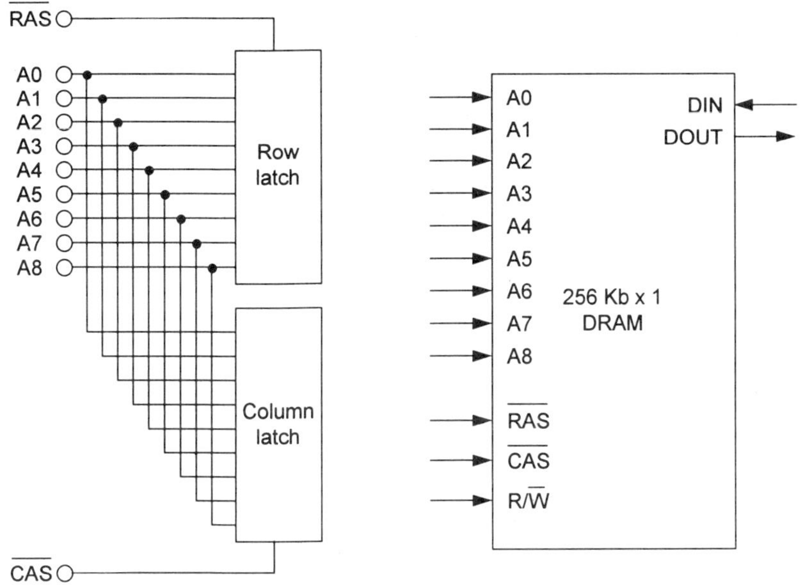

Figure 8.9 Row and column latching.

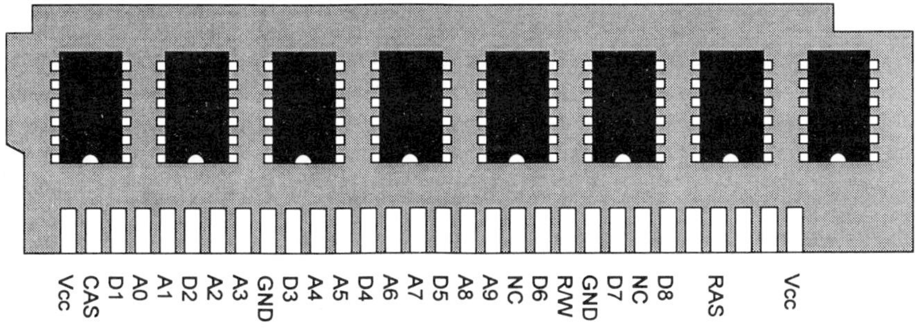

Figure 8.10 30-pin DRAM module (SIMM).

8.7 DRAM memory cell

There is a compromise situation in choosing the capacitance of the DRAM cell. If the capacitance is too large then it takes up a relatively large area and would be relatively slow to charge and discharge. Whereas if the capacitance is too low then the charge stored would leak away too quickly.

One method of fabricating the capacitance is to extend and enlarge the source diffusion area of a transistor. This capacitance results from the junction capacitance between the n-diffusion region and the p-substrate. The actual capacitance values depends on the area of the diffusion and the type of technology (such as 5 μm, 2 μm, and so on). Table 8.2 gives approximate values for the area capacitance for differing technologies. In the case of 5 μm technology the area capacitance is 1×10^{-4} pF/μm^2, whereas for 1.2 μm it is 3.75×10^{-4} pF/μm^2.

Table 8.2 Area capacitance.

	Area capacitance (×10⁻⁴ pF/μm²)		
	5 μm	2 μm	1.2 μm
Gate to channel	4	8	16
Diffusion (when active)	1	1.75	3.75

For example, in the case of 5 μm technology which gives a λ size of 2.5 μm, a transistor with a source diffusion region of 2λ×2λ would have a source capacitance of:

$$C_{Source} = 1\times10^{-4} \times 4\times 2.5\times 2.5 = 0.0025 \text{ pF}$$

This capacitance is far too small to create a large enough capacitance for a DRAM cell as even an area of 16λ^2 (8λ×2λ or 4λ×4λ) would only produce a 0.01 pF capacitance. One solution to increase the capacitance is to place a polysilicon plate over the diffusion area, as illustrated in Figure 8.11. This plate then connects to the supply rail V_{DD} creating a capacitance which is similar to the gate capacitance of a transistor (that is, an area of polysilicon separated by a layer of oxide). Note that the connection between the plate and the supply rail, V_{DD}, has not been included on the diagram so as to aid clarity.

The area capacitance of the plate is 4×10^{-4} pF/μm^2 for 5 μm technology and 16×10^{-4} pF/μm^2 for 5 μm technology. For example, if the polysilicon area is 16λ^2, with 5 μm technology, gives a plate capacitance of:

$$C_{Plate} = 4\times10^{-4} \times 16\times 2.5\times 2.5 = 0.04 \text{ pF}$$

which is four times the equivalent capacitance of a diffusion region. A capacitance of 0.1 pF gives reasonable performance. This requires an area of 40λ^2 (approximately 6λ×7λ). The source capacitance will also add to this figure given a capacitance which is larger than 0.1 pF.

8.8 Sense amplifier

A sense amplifier refreshes a whole row of data. Figure 8.12 shows a basic sense amplifier. Each column has a single sense amplifier. First the bit lines (Bit and $\overline{\text{Bit}}$) are charged to a high level by making the Precharge active HIGH. Next Precharge is made inactive. After this the memory cell is selected and the sense line is made active. If the bit stored was a HIGH then it will be charged up via

Q3 to refresh its level however if it is a LOW then it will be discharged via Q1. The sense amplifier aids the charging and discharging of the cells.

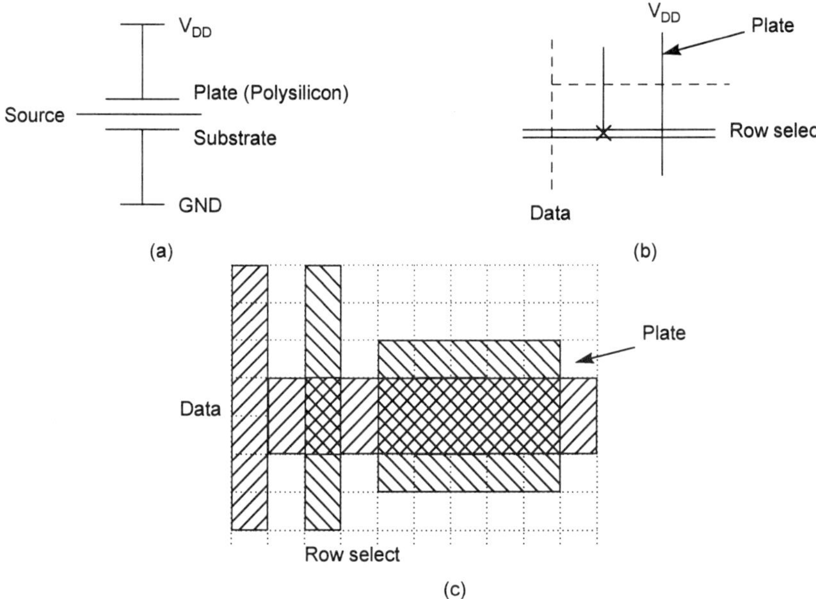

Figure 8.11 DRAM cell.

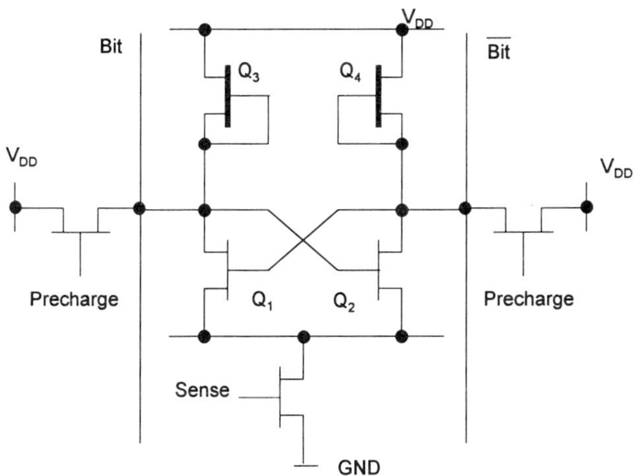

Figure 8.12 Basic sense amplifier circuit.

Figure 8.13 shows the arrangement of the sense amplifier within the memory. Each column has an associated sense amplifier so when a row is selected the whole row will be refreshed. The Column select line only allows a single to reach the IO Bus.

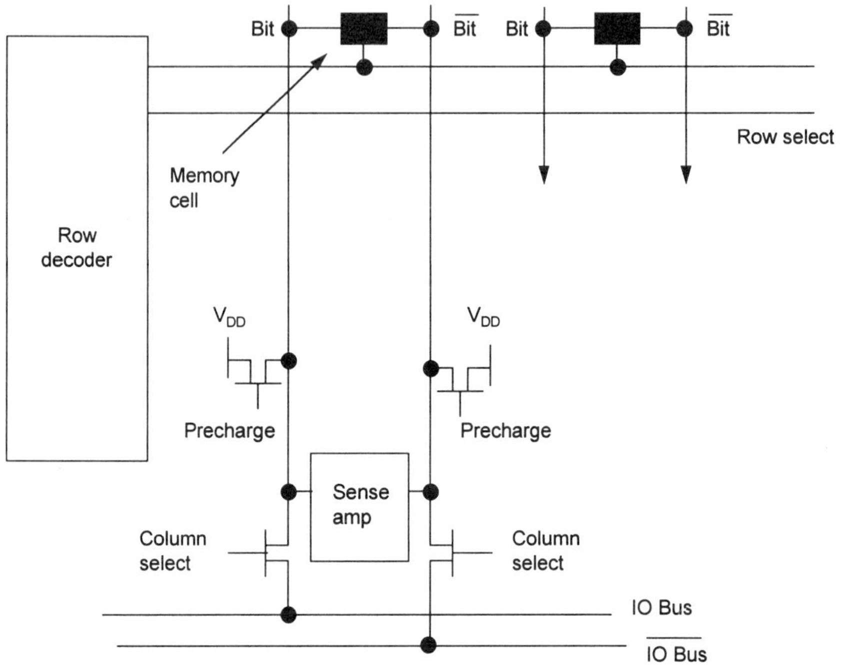

Figure 8.13 Memory arrangement with sense amplifier.

8.9 Memory cache

DRAM is a relatively slow type of memory compared with SRAM. A cache memory can be used to overcome this problem (see Figure 8.14). This is a bank of fast memory (SRAM) that loads data from main memory (typically DRAM). The cache controller guesses the data the processor requires and loads this into the cache memory. If the controller guesses correctly then it is a cache hit, however if it is wrong it is a cache miss. A miss causes the processor to access the memory in the normal way (that is, there may be wait states). Typical cache memory sizes are 16KB, 32KB and 64KB. This should be compared with the size of the RAM on a typical PC which is typically at least 4MB.

Many modern systems have extra cache memory added to improve the hit rate. Typically an 8KB cache memory gives 70% hit rate, a 16KB cache memory 85%, a 32K cache 93% and a 64KB cache 95%. Cache sizes above this do not significantly effect the hit rate and can actually slow the process down as they take so long to fill the cache memory. The Intel 80486 and Pentium processors have built-in cache controllers and have, at least, 8 KB of SRAM cache memory. Intel claim that this has a 96% hit rate, which is an extremely high hit rate for such a small amount of cache memory.

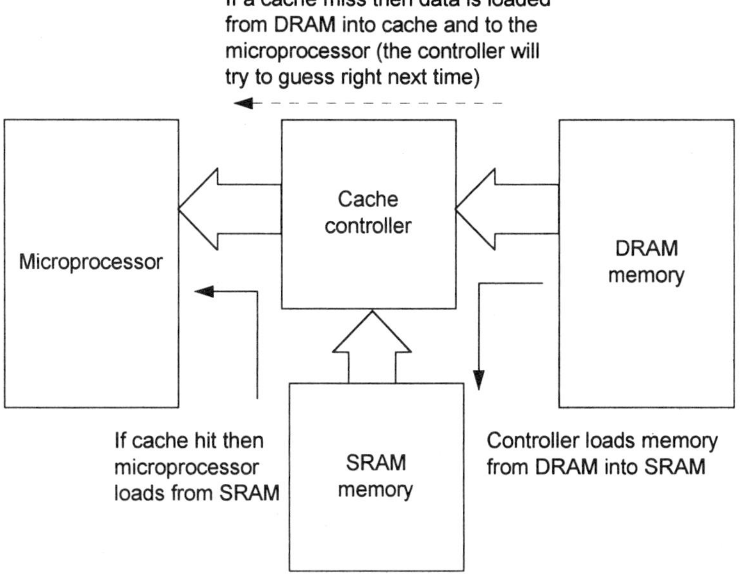

Figure 8.14 Cache operation.

8.10 Exercises

8.1 Discuss the differences between DRAM and SRAM explaining their differing electrical and timing characteristics. Show with the aid of illustrations a typical DRAM and SRAM cell.

8.2 Design a 256 Kb (Kbit) x 1 DRAM IC using four 64 Kbit cells. Describe its architecture and explain how each bit within the memory is addressed.

8.3 For the memory array designed in 8.2 design a basic sense amplifier circuit which can be used to refresh the memory. Explain its operation and how it connects to the array.

8.4 Design a circuit for a 1 Mb (Mbit) × 1 DRAM IC using a number of 64 Kb (Kbit) cells. Describe its architecture and explain how bits within the memory are addressed. Show also how this IC could be used to produce an 8 MB 30-pin SIMM module.

Field Programmable Gate Arrays

9

9.1 Introduction

VLSI devices can now contain many millions of transistors. Full-custom design involves tailor-making the silicon to the required specification. This fine-tuning is normally required when a large amount of transistors are to be fitted onto an IC. It is typical in the design of physically regular structures such as memory arrays and microprocessors. Semi-custom designs use standard cells from stored libraries and the connection between cells is set up when the foundry manufactures the IC. An example of this type of design is shown in the previous chapter where a traffic light controller ASIC was designed. Another type of semi-custom design is a mask-programmed gate array (MPGA) which has a fixed layout of transistors and cells on the silicon and the interconnection between is also set-up when manufactured. This type of design has the advantage in that only the interconnection of the cells has to be designed and stored. Unfortunately, full-custom, semi-custom and MPGA devices are only usually feasible in large production runs and are time consuming to manufacture as they require specialist fabrication facilities. They can also be costly when small modifications or updates are required. Field Programmable Gate Arrays (FPGAs) provide an excellent means of creating a VLSI integrated circuit as they are simply programmed by a FPGA programmer or can be programmed while they are in-circuit. Prototypes can be produced in minutes and are inexpensive in small quantities.

9.2 Programmable devices

The main differences in programmable devices is between:

- mask-programmable and field-programmable;
- erasable and non-erasable.

The mask-programmable types are programmed when they are manufactured whereas the user sets up the field-programmable device with some form of programmer. Mask-programmable devices are expensive in low production runs but are relatively cheap for large production runs, which is opposite for the field-programmable devices. An erasable device allows the stored set-up to be changed whereas the non-erasable type is permanent.

To recap from Chapter 1, programmable logic arrays (PLAs) are made from arrays of AND and OR logic, as illustrated in Figure 1.8, and can be either mask-programmable or field-programmable. They are generally used to implement sets of Boolean equations. If the PLA is mask-programmed then the connections in the arrays are permanently set up when fabricated. When it is field-programmable then the functionality can either be made by fusing links or by electrical programming. Once the functionality has been set up with fusible links then it is not possible to re-program them. When it is electrically programmable then it can be re-programmed as necessary.

Examples of fuse-link programmable devices are programmable read-only memories (PROMs), PLAs, and field programmable arrays (FPGAs). Examples of devices which can be re-programmed are electrically programmable erasable read-only memories (EEPROMs) and electrically programmed ROMS (EPROMs). EEPROMs are erased electrically, while EPROMs require ultra-violet light.

PALs contain arrays of fixed logic functions, typically logic gates and flip-flops. The interconnections between the logic functions are permanent and are either fused or left as they are. In summary, they have:

- fixed array of logic cells;
- fixed connections between cells;
- fusible links to break the connection between cells.

As discussed in Chapter 1, gate arrays are semi-custom ICs with a fixed structure of cells. Routing of one or more layers of metal customizes the design and they differ from PALs in that there are no fixed connections between transistor cells.

Gate arrays contain a large number of transistors laid out in a fixed position with a large number of possible interconnection lines. They also contain general-purpose devices such as input/output structure and oscillators.

In an MPGA all the layers of the device are standard except the two (or more) layers of metal which, once designed, are then sent to an IC foundry for production (there is no need to send the silicon layout as this follows a standard, well-tested, design). With a user programmable gate array, the FPGA, a gate array programmer sets up the required metal layer interconnection by either fusing links, programming switches or with EPROM transistor.

FPGAs were first developed by the Xilinx Company and have since been developed by many other companies, including Actel, Altera, Plessey, AMD and many others.

9.3 FPGA classifications

The are four main classes of FPGAs, these are:

- symmetrical arrays, where the logic blocks are arranged in a grid and the interconnections run horizontally and vertically in between the logic blocks;
- row-based array, where the logic blocks are arranged in rows and the routes interconnect in channels above and below the blocks (like the ASIC design in the previous chapter);
- sea-of-gates, where the logic blocks are laid beside each other and the routes run through them (like a PCB);
- hierarchical PLD, where programmable logic devices (PLDs) interconnect along fixed routes.

The symmetrical and row-based arrays are two of the most popular and are illustrated in Figure 9.1. With the symmetrical array the connections are made between vertical and horizontal metal connection by transistor switches and within programmable switch matrixes. For example, a route may enter from the east of a switching unit and exit to the north, west or south. Several combinations of switching can be set up within the switch, such as one signal entering from the north and exit to the west while another can enter from the west and exit to the south. An example of this is shown in Figure 9.2.

In the row-based approach the interconnection of the logic block is achieved with horizontal lines running though the interconnection channel with routing switches to connect to vertical connections. At various points, vertical lines run through the logic blocks to allow connection between the rows, as illustrated in Figure 9.1.

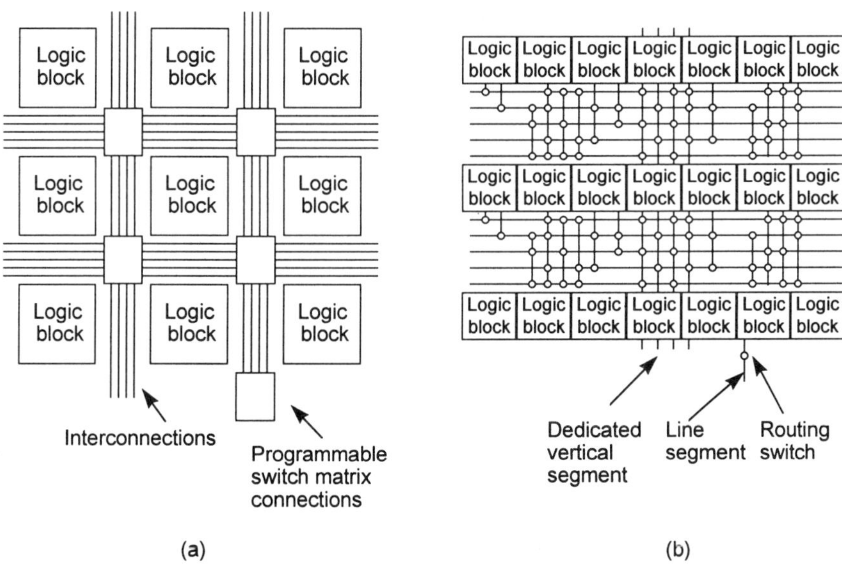

Figure 9.1 (a) Symmetrical and (b) row-based arrays.

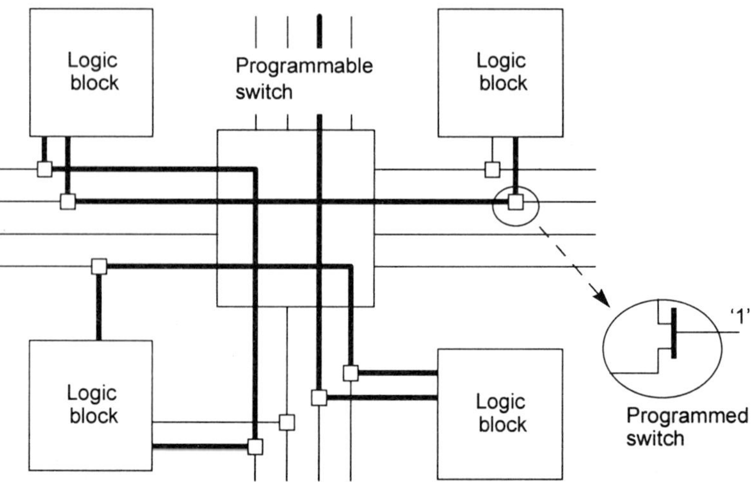

Figure 9.2 Example routing through programmable switch.

9.4 FPGA logic blocks

Each FPGA manufacturer has their own logic block. Most use a look-table approach to generate the required logic. This section looks at two different types used by Actel and Xilinx.

9.4.1 Xilinx XC2000 logic block

A logic block used by Xilinx in their XC2000 series is shown in Figure 9.3. It has four combinational logic inputs: A, B, C and D, a clock input and two outputs: X and Y. A look-up table implements the required combinational logic function. For four input variables it can only output a single logic function (on both X and Y), but for a three input combination it can output two different equations (one on X and the other on Y).

For example, if the required logic function for X is $\overline{\overline{A}.B + \overline{C.D}}$ then the look-up table would contain the following:

```
ABCD X      ABCD X      ABCD X      ABCD X
0000 0      0001 0      0010 0      0011 1
0100 0      0101 0      0110 0      0111 0
1000 0      1001 0      1010 0      1011 1
1100 0      1101 0      1110 0      1111 1
```

The logic block is programmed to set the M_1M_2 bits so that it either uses sequential logic (with a clocked input) or combinational logic (without a clock). If the M_1 bit is a low then the X output will be purely combination because the multiplexer output will be taken from the 0 input of the multiplexer. If the M_1 bit is a high then the input to the multiplexer is taken from the 1 input, which is the output from the D-type latch.

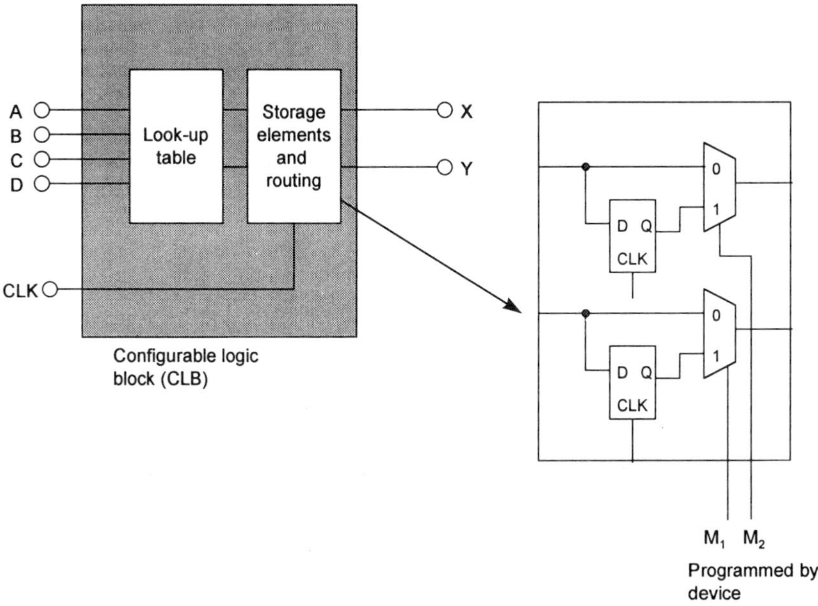

Figure 9.3 Xilinx XC2000-series configurable logic block.

9.4.2 Actel-1 logic module

The Actel logic block (known as a logic module) used in the Act-1 device is simpler than the Xilinx logic block. It consists of three multiplexers and an OR gate, as illustrated in Figure 9.4.

The logic function of a multiplexer with the input *a* on multiplexers input connection 0 and input *b* on multiplexer connection 1, with an input *c* on the address lines is:

$$Output = a.\overline{c} + b.c$$

Thus the logic output for the Act-1 logic block is:

$$Y = \overline{(S_0 + S_1)}.(\overline{S_a}A_1 + S_a.A_0) + (S_0 + S_1).(\overline{S_b}B_0 + S_b.B_1)$$

From this logic function many different logic functions can be generated. Table 9.1 shows a few examples. The Act-2 series uses a similar logic module but has two different programming blocks. The C block contains a similar module to the Act-1 device (for combinational logic) and the S block contain a flip-flops and clocking elements (for sequential logic).

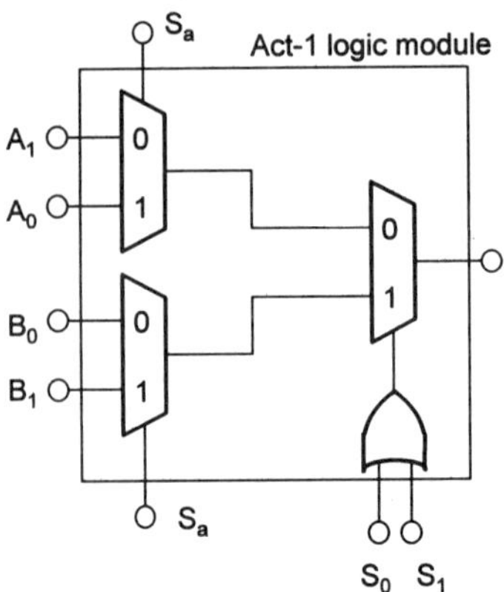

Figure 9.4 Actel Act-1 logic module.

Table 9.1 Example Boolean functions for the Act-1 logic module.

S_0	S_1	A_1	A_0	B_0	B_1	S_a	S_b	Function
0	0	A	B	X	X	C	X	$Y = \overline{C}.A + C.B$
0	0	A	B	X	X	B	X	$Y = \overline{B}.A + B.B = \overline{B}.A + B$
A	B	X	C	D	X	0	0	$Y = \overline{A + B}.C + (A + B).D = \overline{A}.\overline{B}.C + A.D + B.D$
A	B	X	C	0	X	0	0	$Y = \overline{A + B}.C = \overline{A}.\overline{B}.C$
A	B	X	0	C	X	0	0	$Y = (A + B).C = A.C + B.C$
A	B	X	0	1	X	0	0	$Y = (A + B).C = A + B$
A	B	X	1	0	X	0	0	$Y = \overline{A + B} = \overline{A}.\overline{B}$
0	A	B	C	D	E	F	G	$Y = \overline{A}.(\overline{F}.B + \overline{F}.C) + A.(\overline{G}.D + G.E)$
0	A	0	B	C	D	E	F	$Y = \overline{A}.(E.B) + A.(\overline{F}.C + F.D)$
0	A	0	X	B	B	0	C	$Y = A.(\overline{C}.B + C.B) = A.\overline{C}.B + A.C.B$
0	A	0	B	0	0	C	X	$Y = \overline{A}.B.C$

where X represents the don't care state.

9.5 FPGA programming types

The FPGA program sets up the look-up table (if it has one), switching routes and multiplexer addressing to the correct specification. There are three main types of programming for FPGAs, these are:

- Static RAM programming, where the interconnected routes are set up using pass transistors, transmission gates and multiplexers.
- Anti-fuse programming, where the interconnection routes are programmed permanently by devices, known as anti-fuses. These are programmed into a

low-impedance state when a large voltage is applied to them, else they are in a high-impedance state.

- EPROM/EEPROM programming, where transistors are programmed to exist or not. If a transistor exists then it connects in an input to a logic block.

The way that the FPGA is programmed has a great effect on the usage of the device. For example, SRAM programming looses the contents of the program when the power is taken away and the program must thus be reloaded each time the device is powered-up. Anti-fuse types have the disadvantage in that they cannot be re-programmed once they have been programmed. EPROM devices have the advantage that they can be re-programmed but have the disadvantage in that they require a special programmer to re-program them.

9.5.1 Static RAM programming

SRAM programming uses one- or two-way switches and multiplexers to define paths. In the case of the switches with a single NMOS transistor then if the RAM cell associated with that switch is programmed with a 1 then a path will be created for the signal (in one direction). For a two-way switches then a transmission gate can be made from a NMOS transistor in parallel with a PMOS transistor. Figure 9.5 shows three connections between two logic blocks. In this example the top route and the bottom route exists because the SRAM has a logic 1 for these transistors, while the middle route does not exist because the switching transistors gate is set to a logic 0. The diagram also shows a transmission gate which, in this case, passes signals in both directions because its gate is at a logic 1. Note that the look-up table within the logic block (if one exists) is also stored in the SRAM.

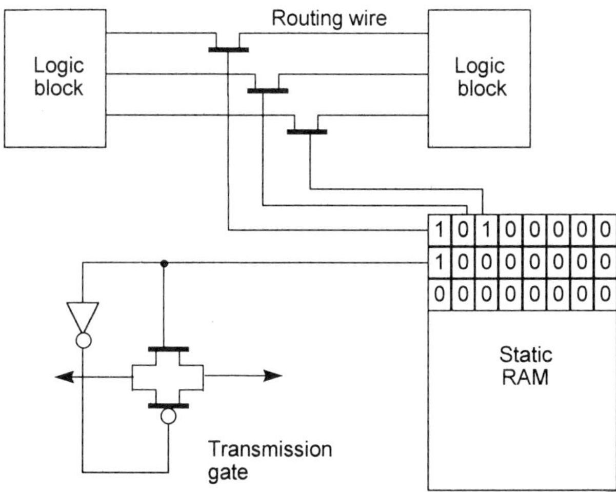

Figure 9.5 Pass transistors and transmission gates in SRAM programming.

The main disadvantage with SRAM programming is that it looses the program stored in the memory when the power is taken away. Thus, some form of non-volatile memory is required to store the program and it also requires some form of program loader. A great advantage with SRAM programming is that there is no need to take the FGPA out of the circuit to re-program it. All that is required is to change the program stored in the non-volatile memory.

SRAM devices also require a relatively large area of silicon because at least five transistors are required for each memory cell of the memory.

9.5.2 Anti-fuse programming

Anti-fuses are used by several manufacturers including the Actel Corporation. An anti-fuse is a high impedance contact until it has a relatively high voltage applied to it which turns it into a low-impedance state. Once fused it cannot be un-fused, thus anti-fuse FPGAs cannot be reprogrammed.

The PLICE anti-fuse is used by Actel and consists of a layer of positively doped silicon (n+ diffusion), a layer of dielectric (Oxygen-Nitrogen-Oxygen) and a layer of polysilicon, as illustrated in Figure 9.6. A relatively high voltage of 18 V is applied across the device. The current that flows (approximately 5 mA) causes the dielectric to melt and form a conductive link between the polysilicon and the diffusion. Two layers of metal connect to the polysilicon and the diffusion, thus causing a low-impedance connection between the two layers of metal. A typical resistance for a fused contact is between 300 and 500 Ω. Note that the reason they are named anti-fuses is that metal fuses normally open-circuit when they are fused, which is opposite to the effect of the anti-fuse.

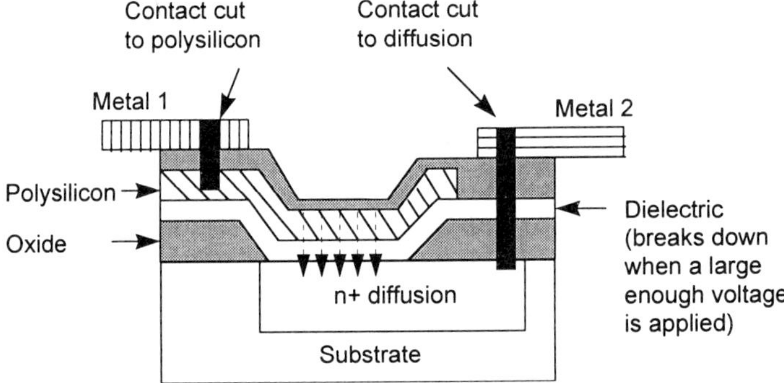

Figure 9.6 PLICE anti-fuse.

9.5.3 EPROM and EEPROM programming

The method used with EPROM/EEPROM programming is the same as the method used in EPROM memories. An EPROM transistor differs from a normal NMOS transistor in that it has two gates, a floating gate and a select gate, as illustrated in Figure 9.7(a). There is no direct electrical connection to the floating gate which is positioned between the select gate and the transistors channel. When no charge exists on the floating gate then the transistor operates in the normal way, which is the un-programmed state. To program the device a relatively large current flows between the drain and source which causes an electrical charge to be trapped under the floating gate. This charge causes the transistor to be permanently off.

An EPROM transistor is re-programmed by removing the trapped charge from the floating gate. This is achieved by exposing the gate to ultra-violet light which excites the trapped electrons. When they gain enough energy they pass through the gate oxide into the substrate.

In the SRAM and anti-fuse types of FPGA, switched connections are made between two metal lines. With EPROM programming the transistors are used as 'pull-down' interfaces to the logic block inputs, as illustrated in Figure 9.7(b). As the figure shows the signal line connects to the select gate of the EPROM transistor. If the transistor has not been programmed in the off state then this acts as an input to the logic block. If it is programmed in the off state then the connected transistor has no effect on the inputs. In the example in Figure 9.7(b) the input logic function is $\overline{B}.\overline{D}$ because the transistors connecting the A and C lines have been programmed in the off state.

Typically many EPROM transistors connect to the same input lines but have different signal lines. The input function to the bit line is a wired-AND function because of the pull-up resistor on the input line. Thus the transistors act both as inputs and implement an AND function. The pull-up resistor unfortunately leads to an increased amount of power dissipation.

An improved device is an EEPROM which is re-programmed electrically and thus can be re-programmed in-circuit.

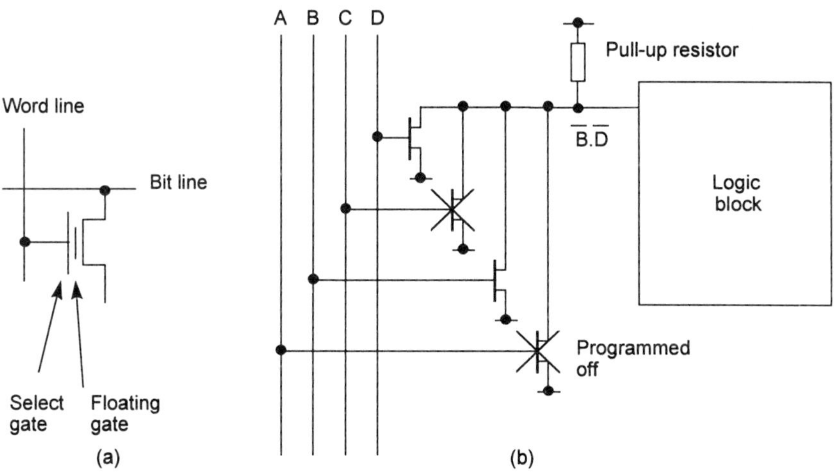

Figure 9.7 (a) EPROM transistors and (b) an example usage of EPROM transistors.

9.6 Practical FPGAs

This section contains information of practical FPGAs.

9.6.1 Xilinx XC-series

Xilinx was the first company to develop FPGAs and are one of the market leaders. All their devices use high-density, low-powered, CMOS. There are currently three generations of devices available, the first generation XC2000-series, the second generation XC3000 and the third generation XC4000-series. The XC2000-series devices are used in this section to illustrate the main concepts because these are easier to illustrate. Differences in the main parameters of the series will be also discussed.

The XC2000, XC3000 and XC4000 device each have SRAM programs and use SRAM cells to save the configurable logic blocks (CLBs) look-up tables and the switch connections. The usage of SRAM means that the FPGA looses its programming once the power is taken away.

Table 9.2 outlines the basic specification for the XC2000 series devices. They have between 64 and 100 CLBs which gives between 1200 and 1800 equivalent gates. The XC2064 is available in either a 48-pin plastic or ceramic dual-in-line (DIL) package or as a 68-pin plastic leaded chip carrier (PLCC). The XC2018 is available in a 68-pin PLCC and as an 84-pin pin grid array (PGA). The DIL package allows the device to be mounted onto one side of a printed circuit board and soldered on the other side, while the PLCC allows the insertion in a PLCC socket. The PGA type allows insertion into a PGA socket, thus allowing it to be easily inserted and to be easily replaced. The XC2000-series CLB was illustrated in Figure 9.3.

Table 9.2 Xilinix XC2000 series.

	XC2064	*XC2018*
CLBs (rows×columns)	64 (8×8)	100 (10×10)
Gates	1200	1800
Logic functions	128	200
Input/outputs	58	74
Program data (bits)	12038	17878

The second generator devices, known as the XC3000 series are outlined in Table 9.3. They have between 2000 and 9000 gates and between 64 and 320 CLBs. The XC3020 has an array of 8×8 CLBs and the XC3030 has an array of 10×10 CLBs.

As with the XC2000-series, the XC3000-series devices are available in a number of packages such as leaded chip carrier (LCC), a pin grid array (PGA) or a quad flat pack (QFP) in either a plastic (P) or a ceramic (C) package. These include the 68-pin PLCC (for the XC3020 and XC3030), an 84-pin PLCC (for the XC3020, XC3030 and XC3042), an 84-pin PGA (for the XC3020, XC3030 and XC3042), a 100-pin PQFP (for the XC3020, XC3030 and XC3042) and a 132-pin PGA (for the XC3042 and XC3064). The QFP type is useful for mounting the devices on surface mounted boards.

The common package types for various different types of device allows for easy upgrade from one device to another. For example, the 132-pin PGA XC3042 can be easily upgraded to a XC3064 by simply replacing the device in the socket with the upgraded version. The pin-outs for devices with the same packaging are identical (although some of the I/O pins may not be connected on the lower specification types).

Table 9.3 Xilinix XC3000 series.

	XC3020	*XC3030*	*XC3042*	*XC3064*	*XC3090*
CLBs (Row×Col.)	64 (8×8)	100 (10×10)	144 (8×8)	224 (16×14)	320 (20×16)
Gates	2000	3000	4200	6400	9000
Logic Functions	128	200	288	488	740
Latches and flip-flops	256	360	480	688	928
Input/outputs	64	80	96	120	144
Program data (bits)	14 779	22 176	30 784	46 064	64 160

9.6.2 XC-2000/XC3000-series architecture

The XC2000- and XC3000-series devices have between 64 and 320 CLBs. These are arranged in a matrix of rows and columns, as illustrated in Figure 9.8. Around the outside of the matrix are input/output blocks (IOBs), and other connections such as a reset pin, clocks, and so on. The IOBs can be programmed as an input or an output buffer. It can also have programmed with options such as the slew rate, CMOS or TTL interfacing, or with tri-state outputs.

Figure 9.8 shows the XC2000-series 8×8 array of CLBs. The first letter of the CLB identifies the row and the second letter the column. For example, the CLB identified by DC is on the fourth row and third column.

In the XC2000-series the switched matrix connects to five main vertical connections that run between the CLBs and four main horizontal connections. There are also two long vertical wires and one long horizontal wire which run alongside the main routing connections. These allow the direct connection of two signals without the requirement to pass through to many switches. This is illustrated in Figure 9.9. In the XC3000-series the switching matrix connection has five vertical and five horizontal connections, as well as three vertical and two horizontal long lines.

The XC2000-series CLBs have four input combination logic connections (A, B, C and D), a clock input (K) and two logic outputs (X and Y). The west side of the CLB connects to the inputs B, C and K, the north to input A, the south to input D and the east to outputs X and Y, as illustrated in Figure 9.9. The rectangular boxes represent a possible programmable switched connection between the horizontal and vertical wire.

It can be seen from Figure 9.9 that there are five possible connections to the D input and six to the A input. Apart from the general purpose and long lines there are also programmable direct connection lines. These are not shown in the Figure and include:

- the connection between the X output and the A input of the CLB directly below;
- the connection between the X output and the D input of the CLB directly above;
- the connection between the Y output and the B input of the CLB directly to the right.

These extra possible connections allow a direct connect of one cell to another without using the switching matrix.

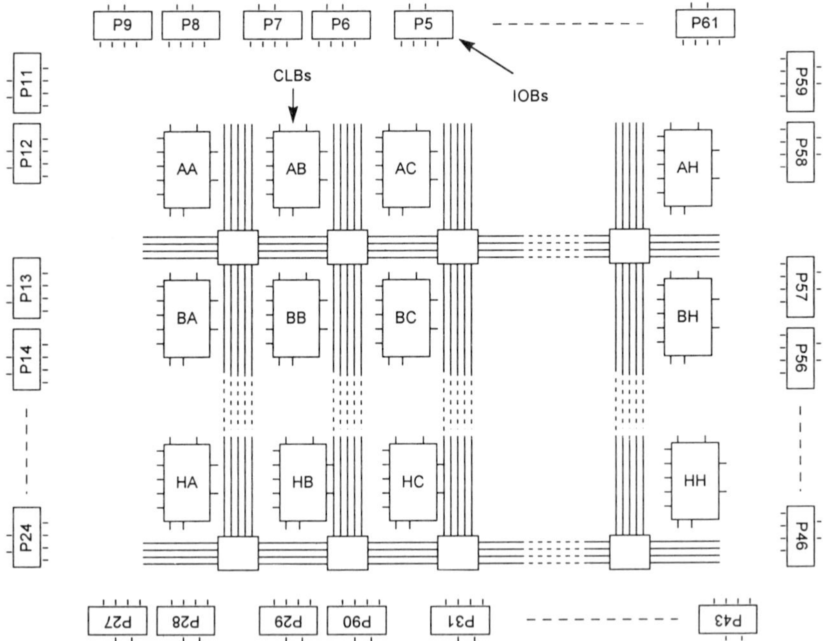

Figure 9.8 XC2000-series architecture.

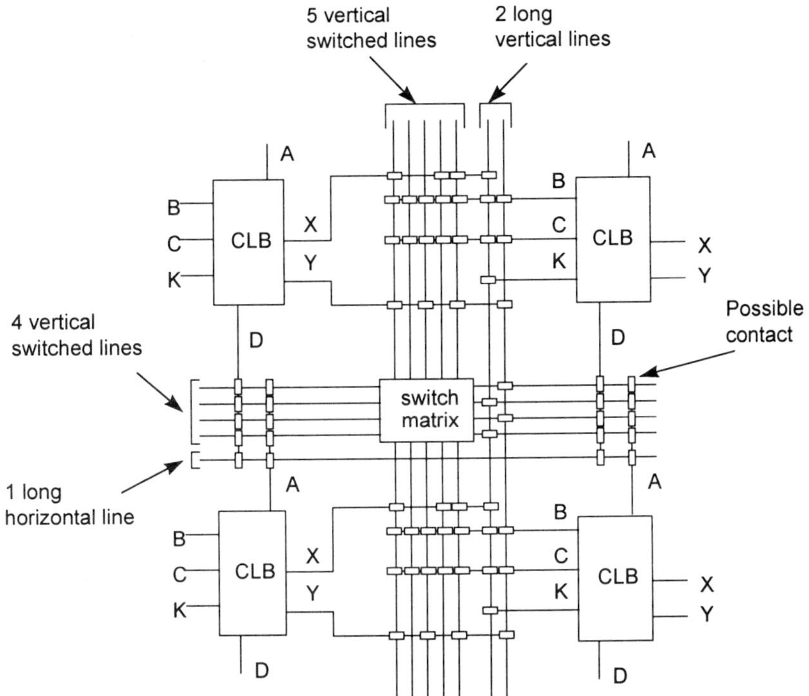

Figure 9.9 Connections to CLBs.

9.6.3 Switching matrix connections

The switching matrix for the XC2000-series devices connects to five vertical and four horizontal connections, as illustrated in Figure 9.10. It contains two smaller switched matrixes, each of which can be programmed with the following possible connections:

- 1 connected to 3, 5, 6, 7 and/or 8;
- 2 connected to 3, 4, 5, 6 and/or 8;
- 3 connected to 1, 2, 4, 7 and/or 8;
- 4 connected to 2, 5, 6, 7 and/or 8;
- 5 connected to 1, 2, 3, 4 and/or 7;
- 6 connected to 1, 2, 4, 7 and/or 8;
- 7 connected to 1, 3, 4, 5 and/or 6;
- 8 connected to 1, 2, 3, 4 and/or 6.

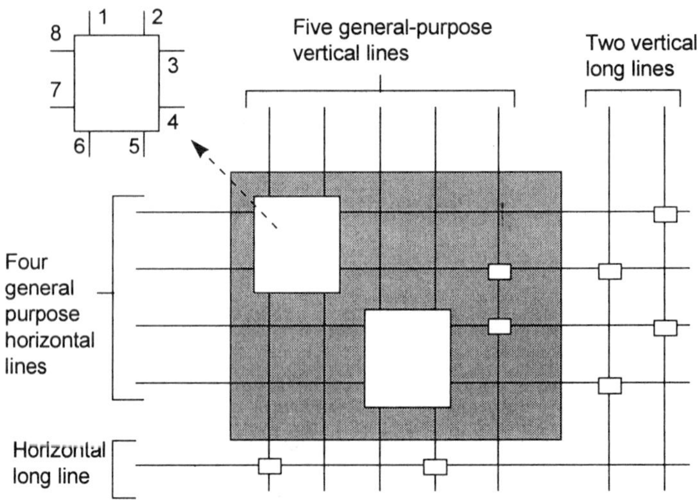

Figure 9.10 XC2000-series switch matrix connections.

The first four vertical general-purpose lines connect to the two switch matrices and the fifth general-purpose line connects to the vertical lines through two switching connections. Figure 9.10 also shows the positions of the programmable interconnections for the horizontal and vertical long lines.

The XC3000-series has a 5×5 switching matrix, the possible connections for connection lines, identified in a clockwise manner, from the top left-hand corner round starting at 1, is:

- 1 connects to 6, 14, 15, 19 and/or 20;
- 2 connects to 6, 7, 13, 14 and/or 15;
- 3 connects to 8, 12, 13, 14, 17 and/or 18;
- 4 connects to 8, 9, 11, 12, 13 and/or 18.

The rest of the switching connections are given in Appendix H.

9.6.4 Configurable logic blocks (CLBs)

The CLB for the XC2000 series is shown in Figure 9.3. The XC3000 and the XC4000 series CLB are more complex than the XC2000-series CLBs, and are illustrated in Appendix H.

9.6.5 Input/output blocks (IOBs)

The IOBs are user-configurable and they provide an interface between the package pin and the internal logic connection. Each XC3000-series IOB has the following:

- TTL or CMOS input threshold levels;
- clocked or direct inputs;
- programmable three-state outputs (on, off or high impedance), each of which can be driven with clocked or direct inputs);
- configurable output inversion;
- configurable slew rate;
- input clamped diodes for electrostatic protection;
- configurable pull-up resistor on inputs.

Figure 9.11 shows an XC3000-series IOB. There are two main inputs from the FPGA array, these are TRO (for three state output) and OUT, and two main outputs to the FPGA array, these are DIN and REGIN (for register in). There are also two clock inputs from the FPGA: INCLK and OUTCLK. The shaded squares in the diagram represent program-controlled memory cells. If the PULL-UP SELECT cell is active then the input has a pull-up resistor connected to it. This resistor helps prevent unused inputs from floating. The TTL OR CMOS SELECT is a programmable cell which allows the input levels to be either TTL or CMOS. The input buffer automatically converts these levels to the required input levels for the FGPA.

The rest of the control cells relate to the output connection to the I/O pad. The SLEW RATE control cells allow the slew rate to be programmed. A fast transition improves critical timing, whereas, slower transitions help to reduce capactitive load currents and minimize system noise. The OUTPUT SELECT control cell allows the either a clock output or a direct output. A tri-state output is selected with he TRI-STATE control cell and an inverted output is selected with the OUT INVERT cell.

There are two input clamped diodes (not shown on the figure). Both are reverse biased when the signal is between the supply rails. One connects to the positive supply rail and the other to GND. If the input fail outwith the supply rails then one of the input diodes conduct and clamps the input voltage of one diode volt drop above or below the supply rail. This helps to protect the device from electrostatic damage and incorrectly connected inputs.

Each output buffer can sink-or-source up to 4 mA of current and can thus drive CMOS or TTL gates.

The XC2000-series IOB is similar to the XC3000-series IOB but does not contain a clocked output (that is, only direct outputs are possible), slewed controls on the output or a configurable pull-up resistor on inputs. It does have control on tri-state output and clock or direct input.

9.6.6 Loading the program into XC-series devices

On power-up the configuration program must be loaded into the SRAM of the FPGA. This can either be done in a serial or a parallel mode. In the parallel mode the data can be loaded from an 8-bit output from an EPROM memory or via a peripheral interface.

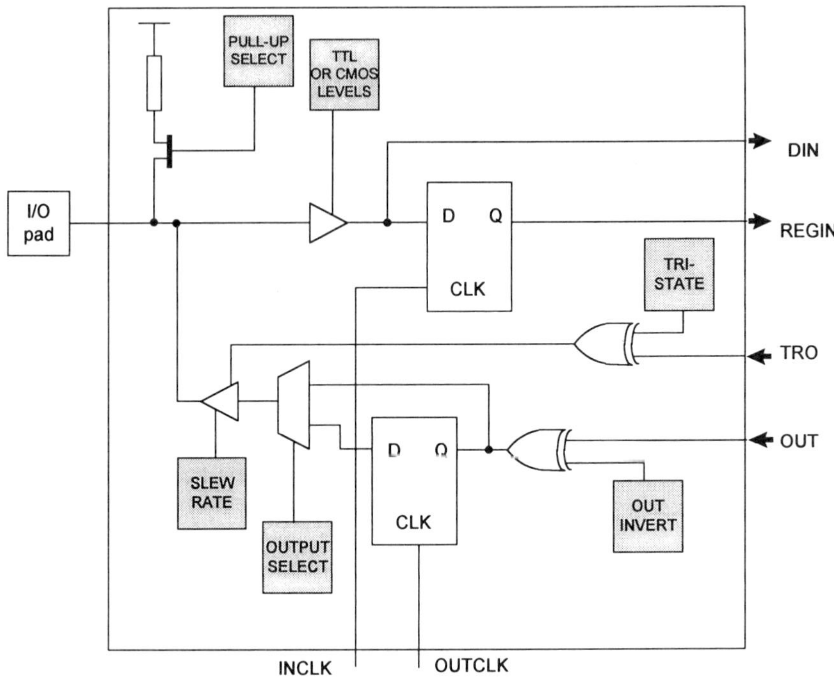

Figure 9.11 XC3000-series input/output block.

The three master bits ($M_1M_2M_3$) allow different modes of programming. To load a program the FPGA can either load its own data in an active mode (that is, the clocking of data is driven by its own internal clock) or it can passively load data (the clocking of data is driven by external device). The data can also either be loaded in a serial manner (one bit at a time) or in a parallel manner (eight bits at a time). Table 9.4 outlines the main modes for the mode bits. For example, if these are set to 000 then the FPGA operates the clock and the data is loaded in a serial mode. In mode 001 the FPGA operates the clock and addresses the external memory from address 0000 up to the end of the address, whereas mode 011 accesses the external memory from the address FFFF down to the end of the address.

In the serial mode one bit is sent at a time. A special IC, known as the XC1736 Serial Configuration PROM, allows the FGPA to load the program contained in the EPROMs memory. Upon power-up, or reset, it operates the CCLK signal line and loaded the data from the DATA line on the serial loader to the DIN line on the FPGA, as illustrated in Figure 9.11. The EPROM within the serial loader can store up to 32 Kbits of data and can be cascaded to load larger programs. Notice that the mode bits are set to 000 which allows the FPGA to actively operate the clock and that the bits are loaded in a serial mode. This mode is known as master serial mode.

Table 9.4 Mode programming bits.

$M_1M_2M_3$	*Clock*	*Mode*	*Data*
000	Active	Master	Serial
001	Active	Master	Parallel (start at address 0000 up)
011	Active	Master	Parallel (start at address FFFF and go down)
101	Passive	Peripheral	Parallel
111	Passive	Slave	Serial

In the master parallel loader the mode bits are set to 001, as illustrated in Figure 9.12. In this mode the FPGA uses some of the I/O pins as address and data lines. At the end of the initialization these revert back to general purpose I/O pins. On initialization the FPGA loads data from the PROM at a rate of one byte at a time from an address starting at 0000. The circuit illustrated in Figure 9.13 shows a 12-bit addressable EPROM which gives 2048 addressable locations. Each location stores 1 byte of data so that the total memory will be 2 KB.

The FPGA can use up to 16 address lines which gives a total addressing capability of 64 KBs (2^{16}) of data.

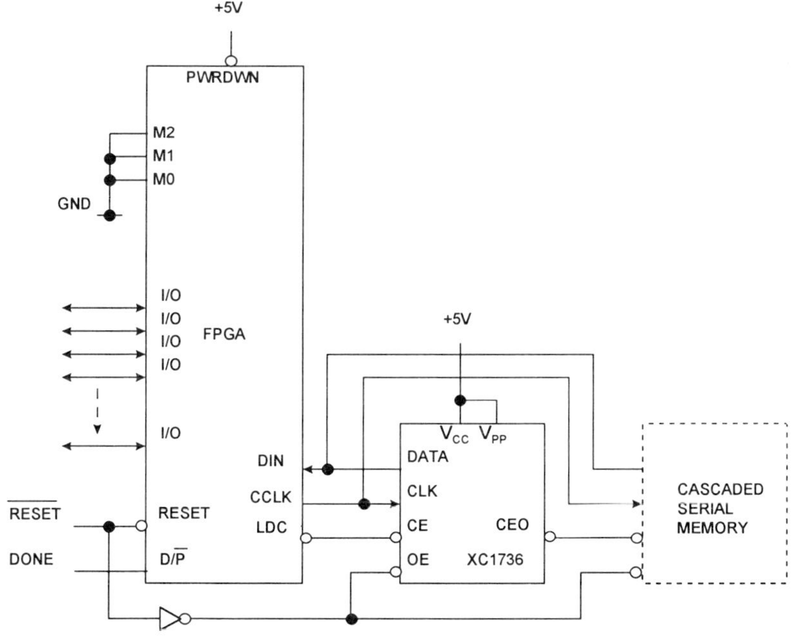

Figure 9.12 Loading the program with master serial mode.

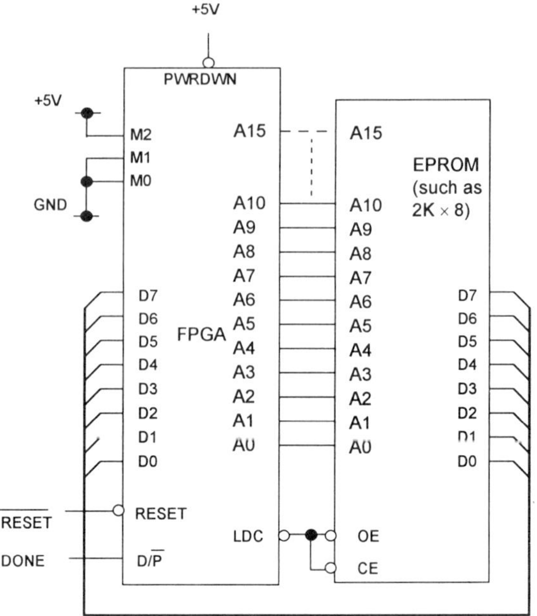

Figure 9.13 Loading the program with master parallel mode.

9.7 Conclusions

FPGAs are extremely useful in implementing functions with a large number of input or output variables. For example if a system has 32-input variables and 8-output variables then an equivalent memory would require a storage of 4 GBs of data and a 64-bit input would require 17 179 869 184 GBs of data.

9.8 Exercises

9.1 Design a circuit, using two AND gates and an inverter, for a 1-bit addressable multiplexer.

9.2 Prove that the output of the multiplexer given Figure 9.4 is

$$Output = a.\bar{c} + b.c$$

and thus prove that the output of the Act-1 logic module, given in Figure 9.4, is:

$$Y = \overline{(S_0 + S_1)}.(\overline{S_a}A_1 + S_a.A_0) + (S_0 + S_1).(\overline{S_b}B_0 + S_b.B_1)$$

9.3 Complete Table 9.5 for the different programming technologies for FPGAs.

Table 9.5 Table for programming types.

Programming type	*Volatile*	*Re-programming*
SRAM	Yes [ ] No [ ]	In-circuit [ ] No re-program [ ] Out-of-circuit [ ]
PLICE anti-fuse	Yes [ ] No [ ]	In-circuit [ ] No re-program [ ] Out-of-circuit [ ]
EPROM	Yes [ ] No [ ]	In-circuit [ ] No re-program [ ] Out-of-circuit [ ]
EEPROM	Yes [ ] No [ ]	In-circuit [ ] No re-program [ ] Out-of-circuit [ ]

9.4 Prove the functions in stated in Table 9.1.

9.5 An Act-1 logic module can be used to implement many different logic functions. For example, referring to Table 9.1, the equation:

$$A.\overline{C}.B + A.C.B$$

can be made to operate as an AND gate if C is set to a 1 (or a 0). Note that for this the S_0 input will be a 0, the S_1 input set to A, A_1 to 0, B_0 to B, B_1 to a B, S_a to 0, S_b to X.

Using this method, design an Act-1 logic module so that it implements the following 2-input gates:

(a) NAND gate; (b) NOR gate;
(c) OR gate; (d) Exclusive-OR gate.

9.6 By setting some inputs to a 0 or a 1, design a circuit using an Act-1 logic module which implements the following functions:

(a) $X.\overline{Z} + X.Z$ (b) $\overline{X}.Z + \overline{Y}.Z$

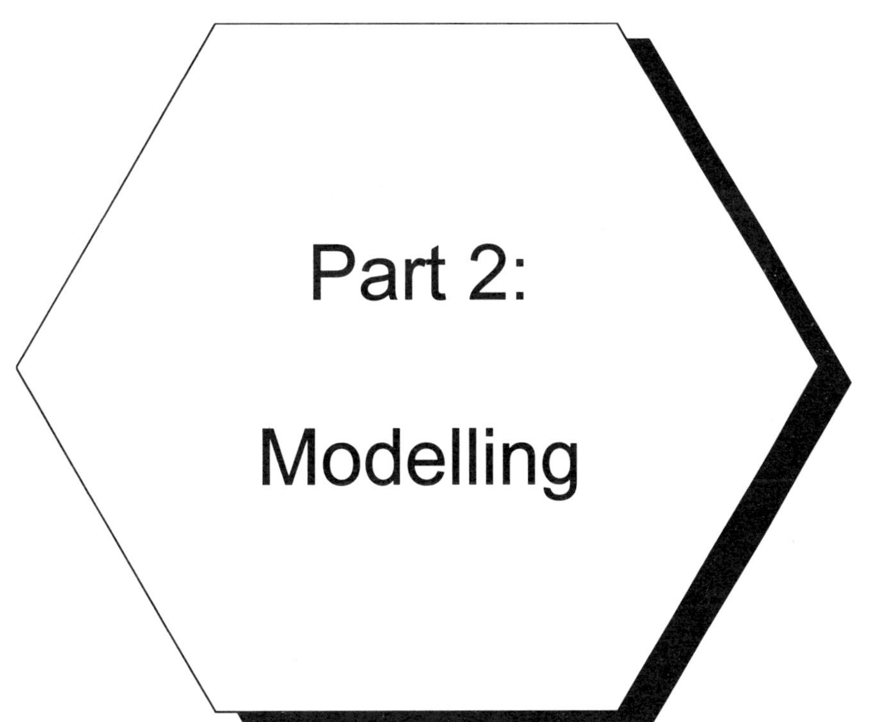
Part 2:
Modelling

High-Level Modelling 10

10.1 Introduction

Modelling and simulation are important in system development. The main objective of simulation is to verify that a design meets the specification. On many occasions the specification itself must be modelled to determine if it meets the requirements. In a large project it is often necessary to model the whole system or part of it before any designs can take place. This modelling allows the designers to learn about the system and make the necessary design decisions. Questions that may be asked might be: Will it work?, Will it get too hot?, Will it be fast enough?, Will it respond to inputs?, Can it handle enough inputs?, Will it be 'idiot-proof'?, and so on.

Modelling and the resultant simulation can involve many different methods, such as:

- high-level language modelling such as ANSI-C or VHDL;
- mathematical modelling;
- prototype (or breadboard) modelling;
- P-Spice modelling for analogue simulations.

10.2 High-level language modelling and emulation

Complex parts of a system can be modelled and/or emulated using a high-level language. One of the most common high-level languages for this purpose is ANSI-C. This type of modelling is useful for several reasons, for example:

- It is relatively easy to specify mathematical and Boolean equations.
- It is relatively easy to model repetition and decision.
- The modelling software can be easily modified.
- Emulation software can be written for the system to be interfaced to.
- It produces portable code that can be run on different systems. Typical standards include ANSI-C (for software portability) and X-Windows (for graphical interfaces).
- Relatively inexpensive development tools are required. Many automated electronic design packages require expensive workstations and a great deal of maintenance.

- Their output can be easily outputted to spreadsheets for analysis.

This chapter discusses how ANSI-C can be used to develop code which models or emulates digital logic systems.

10.3 Modelling combinational logic

In this section a Boolean equation is processed to produce a truth table. The example Boolean equation used is:

$$Z = \overline{(\overline{A+B} + (A.C)).C}$$

Figure 10.1 gives a schematic representation of this Boolean function.

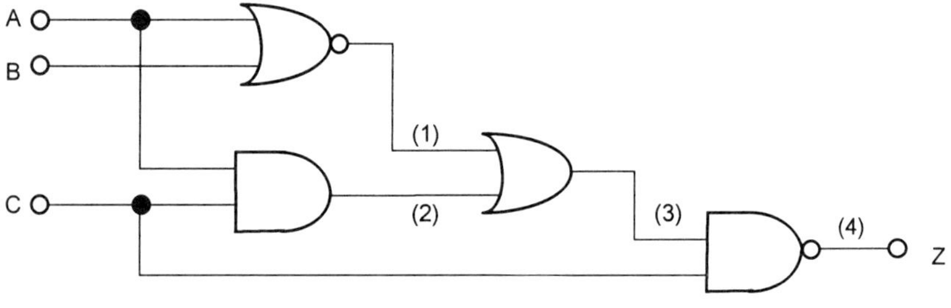

Figure 10.1 Schematic representation of the function $Z = \overline{(\overline{A+B} + (A.C)).C}$.

In order to verify the program it must first be tested against expected results. Table 10.1 gives a truth table showing the logical level at each point in the schematic and Table 10.2 gives the resulting truth table.

The four nodes numbered on this schematic are:

(1) $\overline{A+B}$ (2) $A.C$

(3) $\overline{A+B} + (A.C)$ (4) $\overline{(\overline{A+B} + (A.C)).C}$

Program 10.1 shows that the permutations of the truth table input variables (that is, `000`, `001`, `010`, `011`, ..., `111`) are generated using three nested `for` loops. The inner loop toggles C from a 0 to a 1, the next loop toggles B and the outer loop toggles A. The Boolean functions (AND, NAND, OR, NOR and NOT) use the logical operators `&&` and `||`. In ANSI-C these operators treat a value of 0 (zero) as FALSE and any other value as TRUE.

Test run 10.1 shows a sample run of the program. It should be noticed that the results are identical to the truth table generated by analyzing the schematic.

Table 10.1 Truth table.

A	B	C	$\overline{A+B}$ (1)	$A.C$ (2)	$\overline{A+B}+(A.C)$ (3)	$(\overline{A+B}+(A.C)).C$	$\overline{(\overline{A+B}+(A.C)).C}$ (4)
0	0	0	1	0	1	0	1
0	0	1	1	0	1	1	0
0	1	0	0	0	0	0	1
0	1	1	0	0	0	0	1
1	0	0	0	0	0	0	1
1	0	1	0	1	1	1	0
1	1	0	0	0	0	0	1
1	1	1	0	1	1	1	0

Table 10.2 Truth table.

A	B	C	Z
0	0	0	1
0	0	1	0
0	1	0	1
0	1	1	1
1	0	0	1
1	0	1	0
1	1	0	1
1	1	1	0

Program 10.1

```
#include <stdio.h>
#define  FALSE    0
#define  TRUE     1
   /* ANSI C function prototypes */
int      AND(int x,int y);
int      NAND(int x,int y);
int      NOR(int x,int y);
int      OR (int x,int y);
int      NOT(int x);

int      main(void)
{
int      a,b,c,z;
/* Go through all permutations of truth table */

   puts("     A      B      C      Result");
   puts("    ***************************");

   for (a=FALSE;a<=TRUE;a++)
      for (b=FALSE;b<=TRUE;b++)
         for (c=FALSE;c<=TRUE;c++)
         {
            z=NAND(OR(NOR(a,b),AND(a,c)),c);
            printf("%6d %6d %6d %6d\n",a,b,c,z);
         }
   return(0);
}
```

```
int       AND(int x,int y)
{
   if ( x && y ) return(TRUE);
   else return(FALSE);
}

int       NAND(int x,int y)
{
   if ( x && y ) return(FALSE);
   else return(TRUE);
}

int       OR(int x,int y)
{
   if ( x || y ) return(TRUE);
   else return(FALSE);
}
int       NOR(int x,int y)
{
   if ( x || y ) return(FALSE);
   else return(TRUE);
}
int       NOT(int x)
{
   if (x) return(FALSE);
   else return(TRUE);
}
```

Test run 10.1

```
A      B      C      Result
*************************
0      0      0      1
0      0      1      0
0      1      0      1
0      1      1      1
1      0      0      1
1      0      1      0
1      1      0      1
1      1      1      0
```

It is also possible to use in-line function definitions for the Boolean functions. The statement

```
#define _AND(_x,_y) ( (_x && _y) ? TRUE : FALSE
```

is a shorthand way of representing the `if` statement. If this statement is TRUE, it yields a logical TRUE; else a FALSE. The operators ? and : can be interpreted as 'then' and 'else', respectively.

Program 10.2 is the equivalent of program 10.1 but uses the `#define` directive for the basic Boolean functions. Care must be taken when using the `#define` statements instead of functions as much of the operation of the code conversion is invisible and can thus cause errors which are difficult to trace. It is important that a `#define` substitution is tested independently and then within a program before it can be proved to be functioning correctly.

Program 10.2

```
#include <stdio.h>
#define  FALSE    0
#define  TRUE     1

#define  _AND(_x,_y)        ((_x && _y) ? TRUE   : FALSE)
#define  _OR(_x,_y)         ((_x || _y) ? TRUE   : FALSE)
#define  _NAND(_x,_y)       ((_x && _y) ? FALSE  : TRUE)
#define  _NOR(_x,_y)        ((_x || _y) ? FALSE  : TRUE)
#define  _NOT(_x,_y)        ((_x )      ? FALSE  : TRUE)

int   main(void)
{
int   a,b,c,z;

/* Go through all permutations of truth table */
   puts("     A      B      C      Result");
   puts("     **************************");

   for (a=FALSE;a<=TRUE;a++)
      for (b=FALSE;b<=TRUE;b++)
         for (c=FALSE;c<=TRUE;c++)
         {
            z=_NAND(_OR(_NOR(a,b),_AND(a,c)),c);
            printf("%6d %6d %6d %6d\n",a,b,c,z);
         }
   return(0);
}
```

After the in-line function macros have been fully tested, they can be placed in a user-defined header file. This file should not be placed in the directory which contains the standard header files, but should be inserted into a special user header file directory or placed in the same directory as the source code. In this case, the macros will be put into a header file *boolean.h*. The technique of using header files allows other source code files to access the pre-processor in-line functions, function prototypes or constant definitions.

Where speed of code is important, in-line function definitions are likely to operate faster than the equivalent function as they do not involve parameter passing. Program 10.3 shows the modified code with the included header file. The contents of the file *boolean.h* is shown next; this includes the definition of the Boolean function using macros and also the function declarations.

A user-defined header file is inserted into the program using the `#include " "` pre-processor directive.

File listing: boolean.h

```
#define  FALSE 0
#define  TRUE  1
/*    Definition using macros                                   */
#define  _AND(_x,_y)     ((_x && _y) ? TRUE   : FALSE)
#define  _OR(_x,_y)      ((_x || _y) ? TRUE   : FALSE)
#define  _NAND(_x,_y)    ((_x && _y) ? FALSE  : TRUE)
#define  _NOR(_x,_y)     ((_x || _y) ? FALSE  : TRUE)
#define  _NOT(_x,_y)     ((_x )      ? FALSE  : TRUE)
```

```
/* ANSI C Boolean function prototypes */
int      AND(int x,int y);
int      NAND(int x,int y);
int      NOR(int x,int y);
int      OR (int x,int y);
int      NOT(int x);
```

Program 10.3

```
#include <stdio.h>
#include "boolean.h"

int   main(void)
{
int   a,b,c,z;

/* Go through all permutations of truth table */

   puts("     A      B      C      Result");
   puts("     **************************");

   for (a=FALSE;a<=TRUE;a++)
      for (b=FALSE;b<=TRUE;b++)
         for (c=FALSE;c<=TRUE;c++)
         {
            z=_NAND(_OR(_NOR(a,b),_AND(a,c)),c);
            printf("%6d %6d %6d %6d\n",a,b,c,z);
         }
   return(0);
}
```

Program 10.4 uses bitwise operators to implement the Boolean functions. The single ampersand (`&`) operation is the bitwise AND function and should not be confused with the logical AND statement (`&&`). Similiarly, the verical bar (`|`) operation is a bitwise OR function, which should not be confused with the logical OR function (`||`).

Program 10.4

```
#include <stdio.h>
#include "boolean.h"
int      main(void)
{
int      a,b,c,z;

/* Go through all permutations of truth table */
   puts("     A      B      C      Result");
   puts("     **************************");

   for (a=FALSE;a<=TRUE;a++)
      for (b=FALSE;b<=TRUE;b++)
         for (c=FALSE;c<=TRUE;c++)
         {
            z=NAND(OR(NOR(a,b),AND(a,c)),c);
            printf("%6d %6d %6d %6d\n",a,b,c,z);
         }

   return(0);
}
```

```
int       AND(int x,int y)
{
   return(x&y);
}

int       NAND(int x,int y)
{
   return( !(x&y) );
}

int       OR(int x,int y)
{
   return( x|y );
}

int       NOR(int x,int y)
{
   return ( !(x|y) );
}

int       NOT(int x)
{
   return(!(x));
}
```

10.4 Boolean sequences

Program 10.5 uses a two-dimensional array to set up a 3-bit input sequence applied to a digital circuit. The Boolean equation used is:

$$Z = (in_1.in_2) + in_3$$

Figure 10.2 shows the initialized `input_states[]` array showing groups of binary digits arranged into rows. Initialization of these digits is achieved by grouping them together within braces. Test run 10.2 shows a sample run.

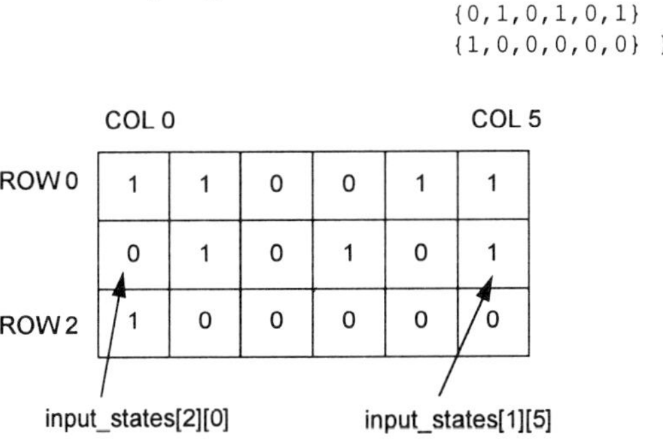

Figure 10.2 Array elements.

Program 10.5

```
/* prog7_8.c      */
#include  <stdio.h>
#define    NO_STATES   6

void    process_states(int nstates,int in[][NO_STATES],int out[]);
void    print_results(int nstates,int in[][NO_STATES],int out[]);

int     main(void)
{
int     input_states[3][NO_STATES]= {
         {1,1,0,0,1,1},{0,1,0,1,0,1},{1,0,0,0,0,0}};
int     output_state[NO_STATES];

   process_states(NO_STATES,input_states,output_state);
   print_results(NO_STATES,input_states,output_state);
   return(0);
}

void    process_states(int nstates,int in[][NO_STATES],int out[])
{
int     i;
   for (i=0;i<nstates;i++)
      out[i]=(in[0][i] & in[1][i]) | in[2][i];
}

void    print_results(int nstates,int in[][NO_STATES],int out[])
{
int     i;
   puts("  A  B  C  Z");
   for (i=0;i<nstates;i++)
      printf("%3d%3d%3d%3d\n",in[0][i],in[1][i],in[2][i],out[i]);
}
```

Test run 10.2

```
  A  B  C  Z
  1  0  1  1
  1  1  0  1
  0  0  0  0
  0  1  0  0
  1  0  0  0
  1  1  0  1
```

10.5 Sequential logic

The example in this section simulates a clocked logic circuit with feedback from the output. An applied sequence of bits is set up and the resultant output determined. The Boolean equation is given by

$$Z_n = \overline{A+B} + (\overline{C+Z_{n-1}})$$

where Z_n is the current output state and Z_{n-1} is the previous output state. Figure 10.3 gives a schematic representation of this equation. Program 10.6 gives a program which simulates the input sequence given in Figure 10.4 and Test run

10.3 shows a sample run.

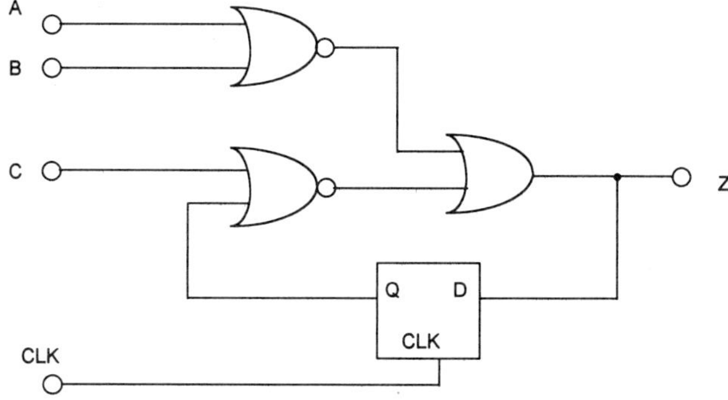

Figure 10.3 Schematic representation of Boolean function.

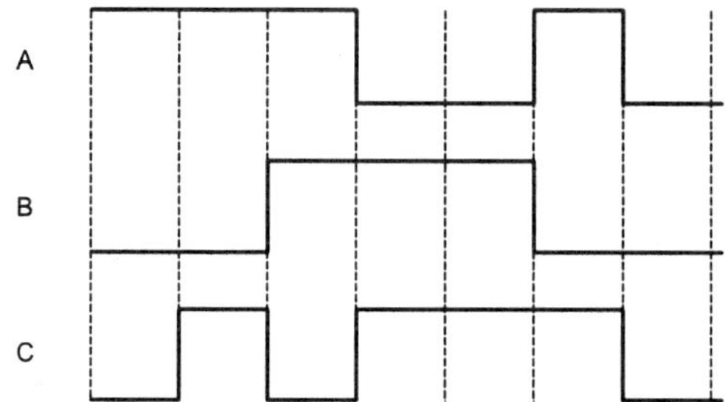

Figure 10.4 Timing diagram.

Program 10.6

```
#include <stdio.h>
#define     NO_STATES      7

void     print_table(int [],int [],int [],int []);
void     process_states(int [],int [],int [],int []);

int      main(void)
{
int      A[NO_STATES]={1,1,1,0,0,1,0};
int      B[NO_STATES]={0,0,1,1,1,0,0};
int      C[NO_STATES]={0,1,0,1,1,1,0};
int      Q[NO_STATES];
   process_states(A,B,C,Q);
   print_table(A,B,C,Q);
   return(0);
}

void     process_states(int a[],int b[],int c[],int q[])
{
int      i;
   q[0]=!(a[0] | b[0]) | (c[0] & 0);      /* initial state Zn-1=0*/
   for (i=1;i<NO_STATES;i++)
      q[i]=!(a[i] | b[i]) | !(c[i] | q[i-1]);
```

```
}

void     print_table(int a[],int b[],int c[],int q[])
{
int      i;
   printf("\n  A  B  C  Q\n");
   for (i=0;i<NO_STATES;i++)
      printf("%3d%3d%3d%3d\n",a[i],b[i],c[i],q[i]);
}
```

Test run 10.3

```
  A  B  C  Q
  1  0  0  0
  1  0  1  0
  1  1  0  1
  0  1  1  0
  0  1  1  0
  1  0  1  0
  0  0  0  1
```

Note that the number of states could be passed as a parameter to the `process_states()` and `print_table()` functions, as shown next.

```
            :          :
   process_states(NO_STATES,A,B,C,Q);
   print_table(NO_STATES,A,B,C,Q);
}

void     process_states(int n,int a[],int b[],
         int c[], int q[])
{
int    i;
       for (i=1;i<n;i++)
          q[i]=!(a[i] | b[i]) | !(c[i] | q[i-1]);
```

10.6 Bit fields

In ANSI C, a structure can be used to define bit fields within an integer (signed or unsigned). These bit fields can vary from 1 to 16 bits wide (or more depending on compiler implementation). The identifier of the bit field is defined after the data type and the width of the bit field is specified after a colon. This is shown in the following example. If there is no bit field identifier then the bits defined in that field are unused. The layout of the bits within the structure `in_out` is given in Figure 10.5.

The bit fields are accessed using the dot operator, such as bit fields 8-10 are accessed by `in_out.motors`. If the bit field type is signed then the most significant bit of the field is a sign flag as signed integers are stored in 2's complement notation. For example, in a 3-bit signed integer bit field a value of –1 (decimal) is stored as `111` (binary), whereas a 3-bit unsigned field `111` (binary) would correspond to 7 (decimal). Thus, the range of a 3-bit signed field is from `011` (3) to `100` (–4) and that of an unsigned 3-bit field from `111` (7) to `000`

(0). Table 10.3 shows the different decimal ranges for a 3-bit binary format.

```
struct
{
   int    limit    :2;   /*Bit fields 0 and 1 used for limit    */
   int    sensors  :2;   /*Bit fields 2 and 3 used for sensors */
   int             :4;   /*Bit fields 4,5,6 and 7 are unused*/
   unsigned int   motors    :3;
                      /*Bit fields 8,9 and 10 are used for motors*/
}  in_out;
```

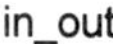

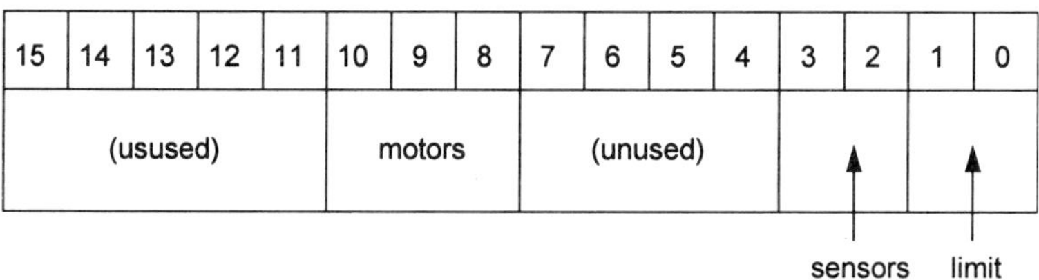

Figure 10.5 Bit field arrangement for `in_out` structure.

Table 10.3 3-bit signed and unsigned decimal values.

3-bit binary format	*Signed decimal value* (`int`)	*Unsigned decimal value* (`unsigned int`)
011	3	3
010	2	2
010	1	1
000	0	0
111	-1	7
110	-2	6
101	-3	5
100	-4	4

Table 10.4 shows an example of a sequence of operations on the bit fields of `in_out`. It is assumed that all the bit fields are set initially to 0.

Table 10.4 Sample operations on `in_out` structure.

Statement	*Bit pattern of* `in_out` *after statement*
`in_out.limit=1;`	000000 000 000 00 01
`in_out.sensors=-2;`	000000 000 000 10 01
`in_out.motors=7;`	000000 111 000 10 01

Program 10.7 uses bit fields to generate a basic traffic light sequence with pedestrian control lights. A structure named `traffic_bit_field` stores the bits for the `RED`, `AMBER` and `GREEN` traffic light status (0 - `OFF` or 1 - `ON`) and the pedestrian status (0 - `DONT WALK`, 1 - `WALK`). The traffic light bit fields are shown in Figure 10.6. Bits 0-2 are used for the traffic light colours and bit 3 for the pedestrian status.

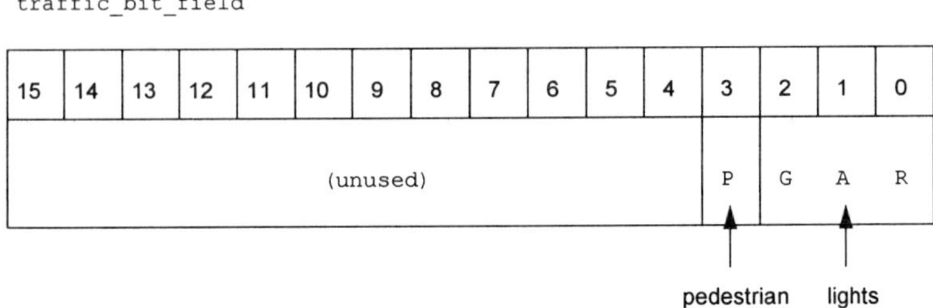

Figure 10.6 Bit field arrangement for `traffic_bit_field` structure.

A function `show_sequence()` is used to mask each of the bits within the structure and display their status. Four bit masks have been set up, these are `RED`, `AMBER` and `GREEN`. The `RED` bit mask operates on the first bit of the lights field, `AMBER` the second and `GREEN` the third. For example, if `traff.lights & RED` is `TRUE` then the `RED` bit is set, else it will be `FALSE` (a 0 is `FALSE` while a `TRUE` is any other value). Table 10.5 shows the required sequence and Test run 10.4 shows a sample run.

Table 10.5 Traffic light and pedestrian control sequence.

Traffic Light	*Pedestrian*
RED	DONT WALK
RED and AMBER	DONT WALK
GREEN	DONT WALK
AMBER	DONT WALK
RED	WALK
RED and AMBER	DONT WALK
GREEN	DONT WALK
and so on.	

Program 10.7

```
#include <stdio.h>

#define     RED       1   /* 1st bit field  */
#define     AMBER     2   /* 2nd bit field  */
#define     GREEN     4   /* 3rd bit field  */
```

```
#define    FALSE   0
#define    TRUE    1

typedef struct
{
   unsigned   int   lights      :3;   /* GAR      */
   unsigned int   pedestrian    :1;   /* P        */
}  traffic_bit_field;   /* XXXX XXXX XXXX PGAR    */

void     show_sequence(traffic_bit_field traf);

int      main(void)
{
traffic_bit_field traffic;
char     ch;

   do
   {
      traffic.lights=RED;
      traffic.pedestrian=FALSE;        /* 0 001 */
      show_sequence(traffic);
      ch=getchar();  /* press any key to continue  */

      traffic.lights=RED+AMBER;           /* 0 011 */
      show_sequence(traffic);
      ch=getchar();  /* press any key to continue  */

      traffic.lights=GREEN;               /* 0 100 */
      show_sequence(traffic);
      ch=getchar();  /* press any key to continue  */

      traffic.lights=AMBER;               /* 0 010 */
      show_sequence(traffic);
      ch=getchar();  /* press any key to continue  */

      traffic.lights=RED;
      traffic.pedestrian=TRUE;            /* 1 001 */
      show_sequence(traffic);
      printf("Press 'x' to exit >> "); ch=getchar();
   }  while (ch!='x');
   return(0);
}

void     show_sequence(traffic_bit_field traff)
{
   if (traff.lights & RED)    printf("RED ");
   if (traff.lights & AMBER)  printf("AMBER ");
   if (traff.lights & GREEN)  printf("GREEN ");

   if (traff.pedestrian==FALSE) puts("<DONT WALK>");
   else puts("<WALK>");
}
```

Test run 10.4

```
RED <DONT WALK>
RED AMBER <DONT WALK>
GREEN <DONT WALK>
AMBER <DONT WALK>
RED <WALK>
Press 'x' to exit >>
```

```
RED <DONT WALK>
RED AMBER <DONT WALK>
GREEN <DONT WALK>
AMBER <DONT WALK>
RED <WALK>
Press 'x' to exit >> x
```

Program 10.8 has an improved structure and uses an array to store the traffic light and pedestrian sequences. Test run 10.5 shows a sample test run.

Program 10.8

```
/*    program to display a traffic light sequence          */
/*    with a pedestrian crossing                           */

#include <stdio.h>

#define     RED      1
#define     AMBER    2
#define     GREEN    4

#define     NO_SEQ   6

#define     FALSE    0
#define     TRUE     1

typedef struct
{
   unsigned int   lights         :3;
   unsigned int   pedestrian     :1;
}  traffic_bit_field;

void      show_sequence(traffic_bit_field traf);

int       main(void)
{
traffic_bit_field    traffic;
char  ch;
int   lights_seq[NO_SEQ]={RED,RED+AMBER,GREEN,AMBER,RED,RED};
int   pedestrian_seq[NO_SEQ]={FALSE,FALSE,FALSE,FALSE,FALSE,TRUE};
int   i;

   puts("Traffic lights sequence program");
   puts("Enter 'x' to exit the program");

   do
   {

      for (i=0;i<NO_SEQ;i++)
      {
         traffic.lights=lights_seq[i];
         traffic.pedestrian=pedestrian_seq[i];
         show_sequence(traffic);
         ch=getchar();
         if (ch=='x') break;
      }

   }  while (ch!='x');
```

```
   return(0);
}

void  show_sequence(traffic_bit_field traff)
{
   if (traff.lights & RED)    printf("RED ");
   if (traff.lights & AMBER)  printf("AMBER ");
   if (traff.lights & GREEN)  printf("GREEN ");

   if (traff.pedestrian==FALSE)  puts("<DONT WALK>");
   else                          puts("<WALK>");
}
```

Test run 10.5

```
RED <DONT WALK>
RED AMBER <DONT WALK>
GREEN <DONT WALK>
AMBER <DONT WALK>
RED <DONT WALK>
RED <WALK>
RED <DONT WALK>
RED AMBER <DONT WALK>
GREEN <DONT WALK>
AMBER <DONT WALK>
RED <DONT WALK>
RED <WALK>
```

10.7 File output

Program 10.9 uses a text file to store the results of a Boolean circuit simulation. The Boolean equation used is $Z=(A.B)+C$. The output file produced is shown in Test run 10.6.

Program 10.9

```
/* Program to determine truth table for boolean        */
/* function Z=(A and B) or C                           */
/* Output truth table is stored to a file              */
/* and then read back from this file                   */

#include <stdio.h>

#define     MAX_STATES     8
#define     STRINGSIZE     100

void     process_eq(int tt[]);
void     dump_table(char fname[],int tt[]);
void     read_table(char fname[]);

int   main(void)
{
int   t_table[MAX_STATES];

   process_eq(t_table);
   dump_table("OUT.DAT",t_table);
   read_table("OUT.DAT");
   return(0);
```

```
}

/* determine truth table */
void     process_eq(int tt[])
{
int      state=0,A,B,C;

   puts("Processing states");

   for (A=0;A<=1;A++)
      for (B=0;B<=1;B++)
         for (C=0;C<=1;C++)
         {
            tt[state]=(A & B) | C;
            state++;
         }
}

/* output truth table to a file */
void     dump_table(char fname[],int tt[])
{
int      state=0,A,B,C;
FILE     *out;

   if ((out=fopen(fname,"w"))==NULL)
   {
      printf("Cannot open %s \n",fname);
      return;
   }

   puts("Writing to file");

   fprintf(out,"A B C Z\n");

   for (A=0;A<=1;A++)
      for (B=0;B<=1;B++)
         for (C=0;C<=1;C++)
         {
            fprintf(out,"%d %d %d %d\n",A,B,C,tt[state]);
            state++;
         }
   fclose(out);
   puts("File written");
}

/* read truth table from a file */
void     read_table(char fname[])
{
int      A,B,C,Z;
FILE     *in;
char     header[STRINGSIZE];

   if ((in=fopen(fname,"r"))==NULL)
   {
      printf("Cannot open %s\n",fname);
      return;
   }

   puts("Reading from file");

   /* get file header ie "A B C Z" */
   fgets(header,STRINGSIZE,in);
```

```
   printf("%s",header);

   while (!feof(in))
   {
      fscanf(in,"%d %d %d %d\n",&A,&B,&C,&Z);
      printf("%d %d %d %d\n",A,B,C,Z);
   }

   fclose(in);
}
```

Test run 10.6

```
Processing states
Writing to file
File written
Reading from file
A B C Z
0 0 0 0
0 0 1 1
0 1 0 0
0 1 1 1
1 0 0 0
1 0 1 1
1 1 0 1
1 1 1 1
```

10.8 Exercises

10.1 Write Boolean logic functions for the following four digital gates:

```
AND3(A,B,C)
OR3(A,B,C)
NAND3(A,B,C)
NOR3(A,B,C)
```

Add these functions to the library of Boolean functions already generated. Refer to Program 10.4.

10.2 Using the functions generated in Question 10.1 determine the truth table for the following equations.

$$Z = \overline{\overline{A + B + C} + (\overline{B}.C.D) + A}$$

$$Z = \overline{\overline{A.C.D}.(\overline{C}.D).B}$$

$$Z = \overline{\overline{A.B.C}.(A + B + C).(\overline{B}.C.D)}$$

10.3 Generate a function which will implement a two input exclusive-OR gate. Test run 10.7 is a sample test run. Note that the exclusive-OR operator in C is a caret (^).

Test run 10.7

```
A  B  Z
0  0  0
0  1  1
1  0  1
1  1  0
```

10.4 Using the function generated in Question 10.3, write a program converts a 4-bit binary code into Gray code. Figure 10.7 gives a schematic of a 4-bit Gray code converter, and Test run 10.8 is a sample run.

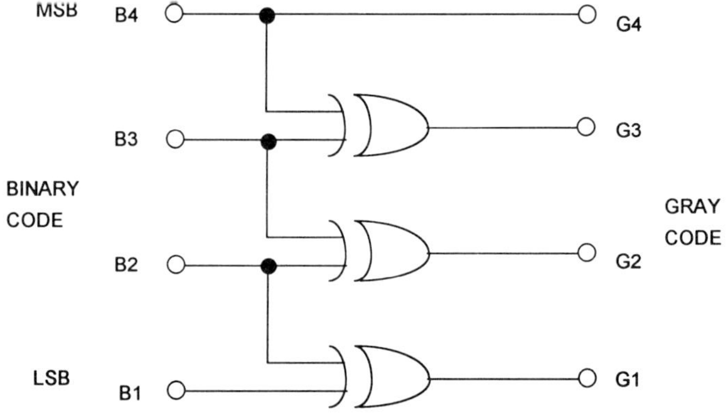

Figure 10.7 Binary to Gray code converter.

Test run 10.8

```
      BINARY                             GRAY
B4      B3      B2      B1        G4      G3      G2      G1
0       0       0       0         0       0       0       0
0       0       0       1         0       0       0       1
0       0       1       0         0       0       1       1
0       0       1       1         0       0       1       0
0       1       0       0         0       1       1       0
0       1       0       1         0       1       1       1
0       1       1       0         0       1       0       1
0       1       1       1         0       1       0       0
1       0       0       0         1       1       0       0
1       0       0       1         1       1       0       1
```

10.5 Write a function which determines the truth table for a 3-input exclusive-OR gate.

10.6 Fault modelling is often conducted using the single stuck-at (ssa) model. This model sets one of the signal lines in the circuit to a given logic level. For the circuit given in Figure 10.8 determine the truth table for the following stuck-at faults:

(a) input A is stuck-at 0;
(b) input A is stuck-at 1;
(c) input B is stuck-at 0;
(d) input B is stuck-at 1;
(e) input C is stuck-at 0;
(f) input C is stuck-at 1.

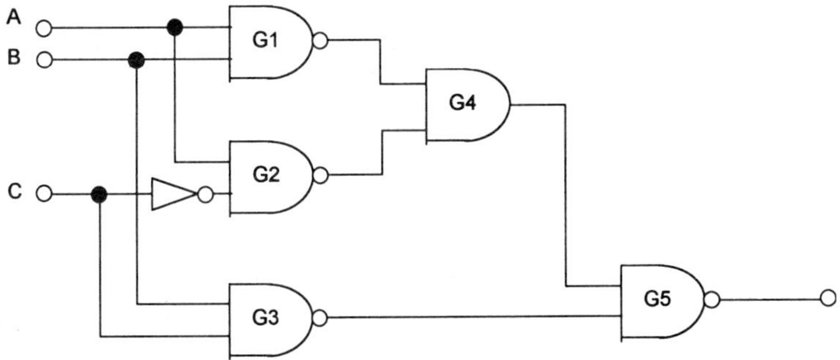

Figure 10.8 Circuit with a single stuck-at fault.

10.7 For the digital circuit given in Figure 10.9 determine the Boolean state table for the given sequence (assume that initially *Z* is 0). Refer to Program 10.6.

Sequence:

```
        start of sequence
        ⇓
A       10011101001110101000
B       00010101100011110100
C       11101010101010101001
```

10.8 Modify the program in Question 10.7 so that the user can enter the A, B and C sequences from the keyboard. Use a function to get the entered sequence. Program 10.10 gives an outline of the program.

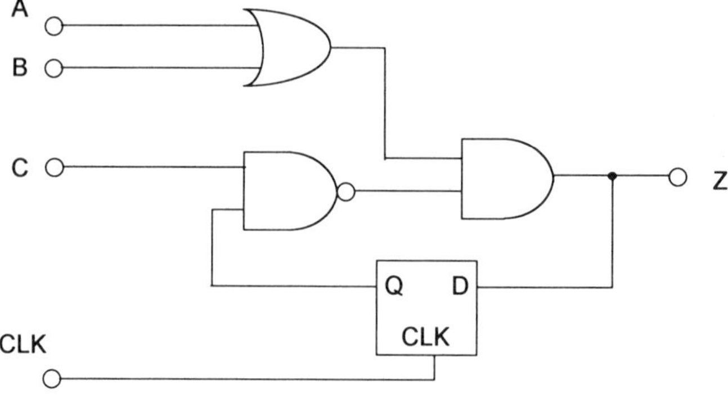

Figure 10.9 Digital circuit.

Program 10.10

```
#define  MAX_STATES  100
           :           :
int      main(void)
{
int   A[MAX_STATES],B[MAX_STATES], C[MAX_STATES], Q[MAX_STATES
   get_sequence(&n_sequence,A,B,C);
   process_sequence(A,B,C,Q);
   print_sequence(A,B,C,Q);
   return(0);
}

void     get_sequence(....)
{
   get_seq(n,"Enter A sequence",a);
   get_seq(n,"Enter B sequence",b);
   get_seq(n,"Enter C sequence",c);
}
   :   :   :   :
   :   :   :   :
```

10.9 Modify Program 10.8 so that it displays the sequence of two traffic lights with a single pedestrian control. Table 10.6 shows a sample sequence and Figure 10.10 shows a possible bit pattern for the structure.

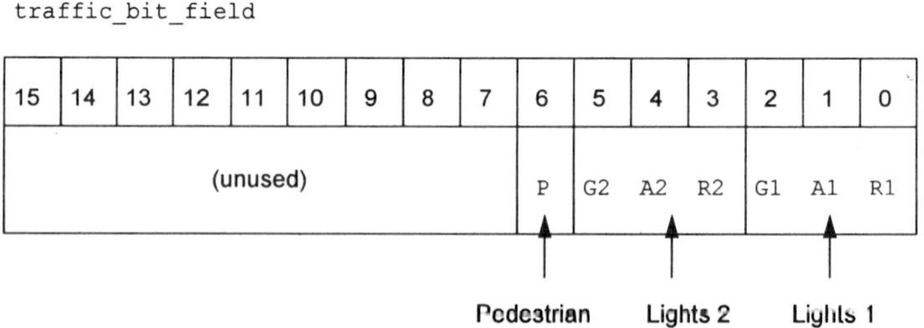

Figure 10.10 Bit field arrangement for `traffic_bit_field` structure.

Table 10.6 Traffic lights and pedestrian control sequence.

Traffic light 1	*Traffic light 2*	*Pedestrian*
RED	RED	DONT WALK
RED	RED and AMBER	DONT WALK
RED	GREEN	DONT WALK
RED	AMBER	DONT WALK
RED	RED	WALK
RED and AMBER	RED	DONT WALK
GREEN	RED	DONT WALK
AMBER	RED	DONT WALK
RED	RED	WALK
RED	RED and AMBER	DONT WALK
	and so on.	

VHDL Modelling 11

11.1 Introduction

The VHSIC (Very High Speed Integrated Circuit) Hardware Description Language (VHDL) is an industry standard language which describes hardware. Many electronic design systems are now standardizing on VHDL as the input and output for simulation, synthesis, layout tools and so on.

11.2 History of VHDL

In the late 1970s and early 1980s, the Department of Defense in the USA funded the development of the VHSIC program. Its main objective was to produce the next generation of integrated circuits. Part of the programs aims was to enhance every part of the design process. Traditional design methods which used schematics were found to be rather lacking in the development of complicated systems. For this purpose VHDL was proposed. The two main aims were:

- to describe complex circuits;
- to standardize on a language which allow designs to be transferred between different systems.

In 1986 the IEEE proposed VHDL as a standard language. After a few revisions it was eventually adopted in 1987 as the IEEE 1076 standard. It is sometimes known as VHDL-87, which is used in the this chapter. An update to the language, VHDL-93 contains many enhancements, but for cross-capability is it best, wherever possible, to use the VHDL-87 standard.

11.3 Structural decomposition

Formal design methods often use a top-down approach where the problem splits into smaller sub-units. Figure 11.1 recaps from Chapter 1 the main layers in the design abstraction. Structural decomposition initially involves the defining the functionality of the system and how it interfaces with other systems. All inputs and outputs are identified and characterized. Next, at the functional layer, the system splits down into a number of discrete blocks. Each of these is either modelled as an algorithm or as a functional block with defined inputs and outputs. Next, at the circuit level, symbols joined by electrical connection represents each block. If specified by an algorithm then functional statements represent each block, or module. Below this layer, electrical symbols and connections (or algorithms) are represented pictorially by transistors and electrical routes. Finally, at the physical level the devices and connections laid out in silicon. This involves the generation of several masks to define the IC layout. These levels are illustrated in Figure 11.2.

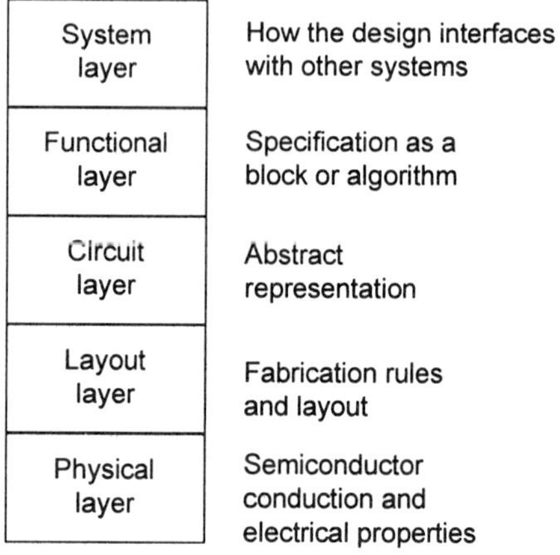

Figure 11.1 Design structure.

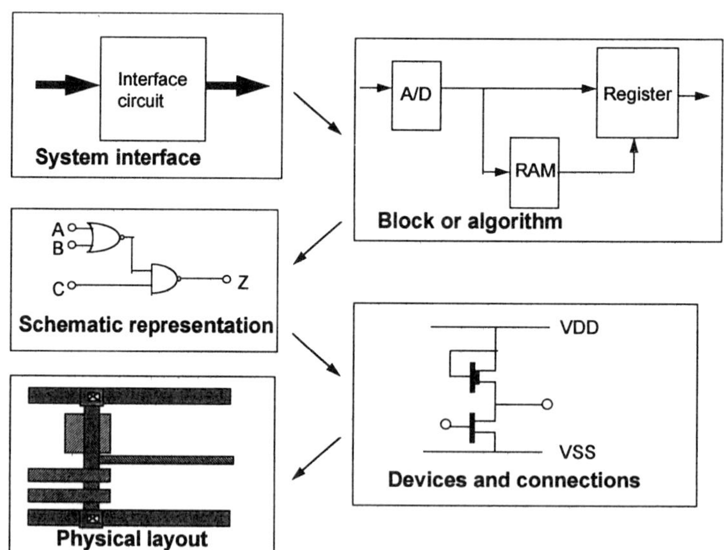

Figure 11.2 Design abstraction.

11.4 Terminology

VHDL is a very powerful hardware description language which can model a simple logic gate or a large complex system. An entity is a hardware abstraction of an actual hardware device. It has a defined input and output. To describe an entity, VHDL uses five main types of constructs, called design units, these are:

- entity declaration;
- architecture body;
- configuration declaration;
- package declaration;
- package body.

11.5 Entities

Typical design methods in software engineering use structural decomposition, where the system is split into a number of modules. These modules can be represented as a structure chart, as illustrated in Figure 11.3. Each module has a standard interface for the parameters passed into and out of it. An electronic system can also be represented with structural decomposition. VHDL uses the term entity for the equivalent of modules in software. Figure 11.4 shows the top-down design for an electronic system, each hatched line block represents a distinct entity with a well-defined interface. The entities can contain functional elements or other entities can be called. If possible, the actual operation of each entity should be independent of all other entities. An entity can be as large as a complicated microprocessor or a simple as a single transistor.

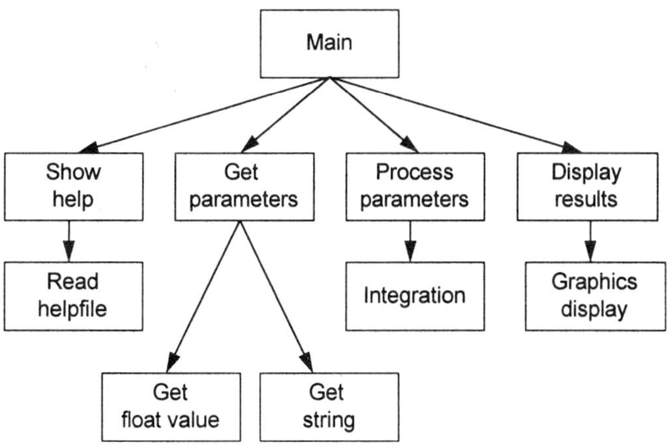

Figure 11.3 Structure chart.

An entity is defined with its input and output interface. For example, VHDL block 11.1 shows the entity statement for a logic block named `LBLOCK` which has three inputs: `IN1`, `IN2` and `IN3`, and three two inputs: `OUT1` and `OUT2`. It is common in the entity statement to insert some comments which give an outline of its functionality. These are inserted after the `--` symbols. Note that the block words in the VHDL block are reserved words in VHDL. The `in` word defines an input port, the `out` defines an output port. A `bit` word identifies that the signal is either a '0' or a '1'.

VHDL block 11.1

```
entity  LBLOCK is
   port(IN1, IN2, IN3: in bit; OUT1, OUT2: out bit);
   -- This is a comment. Three inputs IN1, IN2 and IN3
   -- and two outputs OUT1 and OUT2
end     LBLOCK;
```

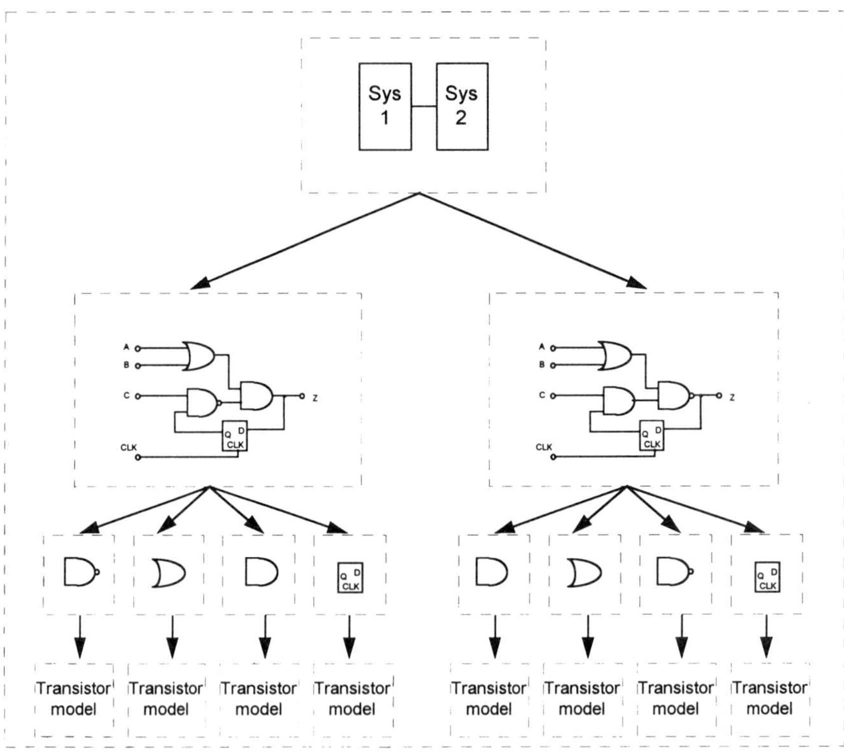

Figure 11.4 Structural design using design entities.

11.6 Architectural bodies

The entity statement defines the interface of an entity, while an architectural body defines the behaviour of an entity. VHDL block 11.2 shows an architectural block definition, named DEF_BLK, for an entity LBLOCK for which the output OUT1 is IN1 or'ed with IN2 and OUT2 is IN2 and'ed with IN3. The reason that the architecture has a distinct name (in this case DEF_BLK) is that a single entity can have one or more architecture to describe its behaviour.

The signal assignment operator <= is similar to the assignment operator (:=) but involves a time delay, that is, the value is assignment at some future time. In the case of VHDL block 11.2 there is no time delay on the operations. An example of the signal assignment with a time delay is:

```
x<=y and z delay 15 ns;
```

which assigns the value of x and'ed with y after a time delay of 15 ns.

VHDL block 11.2

```
entity  LBLOCK is
   port(IN1, IN2, IN3: in bit; OUT1, OUT2: out bit);
   -- OUT1 is IN1 or'ed with IN2, and
   -- OUT2 is IN2 and'ed with IN3
end     LBLOCK;

architecture DEF_BLK of LBLOCK  is
begin
   OUT2 <= IN2 and IN3;
end architecture  DEF_BLK;
```

11.7 VHDL operators

VHDL has a varied set of operators, each with their own precedence. Table 11.1 outlines these. Most of these are obvious in their operation.

Table 11.1 VHDL operators.

Operator class	*Operator*	*Operator*	*Precedence*
Miscellaneous	**	Exponentiation	HIGHEST
	ABS	Absolute value	
	NOT	Negated value	
Multiplying	*	Multiply	
	/	Divide	
	MOD	Modulus	
	REM	Remainder	
Signed	+	Identify	
	-	Negation	

Table 11.1 (Cont.).

Operator class	*Operator*	*Operator*	*Precedence*
Adding/subtraction	+	Addition	
	-	Subtraction	
	&	Concatenation	
Relationship	=	Equal-to	
	/=	Not equal-to	
	<	Less than	
	<=	Less than or equal to	
	>	Greater than	
	>=	Greater than or equal to	
Logical	AND	AND operator	LOWEST
	OR	OR operator	
	NAND	NAND operator	
	NOR	NOR operator	
	XOR	Exclusive-OR operator	

11.8 Identifiers

Identifiers are named items in a VHDL model. These should be as descriptive as possible and must:

- only contain alphabets characters, decimal digits ('0' to '9') and the underline character ('_');
- start with an alphabet character;
- not end in an underline character;
- not contain two successive underline characters.

Examples of identifiers are:

```
COM_BLOCK    integrator_module  led10
```

Examples of invalid identifiers are:

```
output$1     -- invalid character ($)
1invert      -- invalid starting character
data__bus    -- contains two successive underlines
```

Note that the case of characters has no significance, thus the identifier `INVERT1` is the same as `Invert1`.

11.9 Reserved words

As with other high-level languages there are a number of reserved words, or keywords. These reserved words cannot be used as identifiers.

```
abs            access       after       alias       all         and
architecture   array        assert      attribute   begin       block
body           buffer       bus         case        component   configuration
constant       disconnect   downto      else        elsif       end
entity         exit         file        for         function    generate
generic        guarded      if          in          inout       is
label          library      linkage     loop        mp          mod
nand           new          next        nor         not         null
of             on           open        or          others      out
package        port         procedure   process     range       record
select         severity     signal      shared      subtype     then
to             transport    type        units       until       use
variable       wait         when        while       with        xor
```

11.10 Numbers, characters and strings

11.10.1 Numbers

VHDL uses either real or integer numbers. An integer number is a whole number and does not have a decimal point. Large and small numbers are normally represented using the exponent format. Examples of this type are:

```
1e-6       -- 0.000001
5.4e9      -- 5400000000
6.2e-12    -- 0.0000000000062
```

Numbers can also be represented in any base from 2 to 16 by surrounding the number is the hash ('#') character, preceded by the base value. For example:

```
2#101000110#     -- binary, 101000110
16#1f0#          -- hexadecimal, 1F0
8#771#           -- octal, 771
```

11.10.2 Characters and strings

A single character is enclosed by single quotation marks and a string of characters is enclosed in double quotation marks. Examples of uses are:

```
'Z'          -- single character
"Inverter"   -- string
```

A single quotation mark character uses three consecutive single quotation marks and a double quotation mark. For example:

```
'''                              -- single quotation character
"This is an ""example"""         -- This is an "example"
```

Strings can be concatenated (joined together) using the & operator. For example:

```
"This is an example of a"
&"concatenated string"
```

11.10.3 Bit strings

Apart from representing binary and hexadecimal number with their number base there is an easier format for these types. A preceding B in front of a binary string represents a binary value and an X for a hexadecimal number. For example:

```
B"0101001010"     -- binary value
X"1FD20"          -- hexadecimal value
```

Hexadecimal values are useful when representing binary values as they reduce the number of digits by around one-quarter.

11.11 Constants and variables

VHDL, as most high-level software languages, allows the definition constants and variablcs within a modcl. A variablc can takc on any valuc in thc rangc of its type whereas a constant has a fixed value. The definition of constants is similar to other high-level languages where the := assignment operator assigns a value to the constant. It differs from other high-level languages in that the datatype of the constant is defined after the constants are named. Typical examples of constants are:

```
constant speed_of_light:real:=3e8;
constant bits_per_byte:integer:=8;
constant e:real:=2.71828;
constant pi:real:=3.14159;
```

Variables are declared in a similar manner. For example:

```
variable input_value:real;
variable data_out,data_in:integer;
variable x,y,z:real;
```

These constants and variables can appear anywhere within a VHDL model and are assigned values using the assignment operator, such as:

```
input_value:=output_value*2;
counter:=counter+1;
```

The signal assignment operator (<=) should not be confused with the assignment operator (:=). A signal assignment allows for a signal to be updated at a future time period whereas the assignment operator is implemented. For example in the two statements below, y is assigned a value of 10 and z is assigned a value of 10 after a period of 10 ns.

```
y:=10;
z<=10 after 10 ns;
```

11.12 Scalar types

VHDL is a strongly type-oriented language. Each object has a defined type and this type defines how the bits are stored within the object and how it is operated on.

11.12.1 Type declarations

VHDL allows for the declaration of user-defined types. Two examples are given next. The first, byte_range, declares a type in the range 0 to 255, and the second, signed_byte_range, defines a range from –128 to 127. The byte_range type is used for the variable val1 and val2, and the signed_byte_range is used for the variables i and j.

```
type    byte_range is range 0 to 255;
type    signed_byte_range is range -128 to 127;

variable val1,val2: byte_range;
variable i,j:signed_byte_range;
```

The initial values of variables declared within ranges is always the low range values. So in the previous example the initial values of val1 and val2 will be 0 and the initial values of i and j will be –128.

It is also possible to define a range which relates to a constant. For example:

```
constant bits_per_longword:=32;
type long_word_bits is range 0 to bits_per_longword-1;
```

There are a few standard pre-defined types, these include boolean, bit, integer, real and time. These will be covered in the following sections.

11.12.2 Integers and real types

The range of the real type depends on the VHDL implementation but will be stored with at least 32 bits which gives a range of at least –1e38 and +1e38. An integer type range also depends on the implementation but will be stored with at least 32 bits. This gives a range of at least of –2,147,483,647 to +2,147,483,647.

11.12.3 Time and defining units

VHDL has **pre-defined** units for time, these are:

```
1 fs    -- 1e-15 s
1 ps    -- 1000 fs
1 ns    -- 1000 ps
1 us    -- 1000 ns
```

```
1 ms    -- 1000 us
1 sec   -- 1000 ms
1 min   -- 60 sec
1 hr    -- 60 min
```

A few examples are:

```
time_step:=12.5 us;
x<=x1+y1 after 20 ns;
```

The lowest element of time in a simulation is 1 fs. A value of time lower than this is treated as a zero time. The range of time depends on the VHDL implementation but will at least use 32-bit floating point values, giving a range of $-(2^{32}-1)$ to $+(2^{31}-1)$

Time is the only pre-defined unit in VHDL but other user-defined units can be defined using the `units` statement. For example, if a type resistance is to range from 0 to 100 MΩ and uses the units Ω, kΩ and MΩ then the following type declaration can be made:

```
type resistance  is range 0 to 1e8
  units
    ohm;
    kohm=1000 ohm;
    Mohm=1000 kohm;
  end units resistance;
```

The values of the type are expressed as integer multiples of the base unit or distance:

```
type length is range 0 to 1e8
  units
    um;
    mm=1000 um;
    m=1000 mm;
    mil=254 um;
    inch=1000 mil;
  end units length;
```

11.12.4 Enumeration types

Enumerator types allows variables to be limited in the values they can be assigned. For example:

```
type hex_digit is ('0',  '1',  '2',  '3',  '4',  '5',
   '6',  '7',  '8',  '9',  'A',  'B',  'C',  'D',  'E',  'F'),
type bit_type  is  ('0',  '1');
```

```
type range_setting is (high,medium,low);
variable range:range_setting;
variable digit:hex_digit;
```

There are three standard enumerators defined, these are `bit`, `boolean` and `character`. These are discussed in the next three sections.

11.12.5 Boolean

The Boolean data type is a standard pre-defined type and is enumerated with TRUE and FALSE. It is defined as:

```
type boolean  is (FALSE,TRUE);
```

An example of its use is:

```
variable input:boolean;

  input:=true;
```

Boolean operations are either relationship or logical. The main Boolean relationship operators are = (is equal to), /= (is not equal to), <= (less than or equal to), >= (greater than or equal to), < (less than) and > (greater than).

The logical operators also operate on Boolean values, these are **and**, **or**, **nor**, **nand**, **xor** and **not**. Table 11.2 lists the results from these operations.

Table 11.2 Boolean logical operations.

A	*B*	*A and B*	*A or B*	*A nand B*	*A nor B*	*A xor B*	*not A*
False	False	False	False	True	True	False	True
False	True	False	True	True	False	True	True
True	False	False	True	True	False	True	False
True	True	True	True	False	False	False	False

11.12.6 Bit

The bit data type is a standard pre-defined type and is enumerated with the characters '0' and '1'. It is defined as:

```
type bit  is ('0', '1');
```

An example of its use is:

```
variable bit_1:bit;

  bit_1<='1' after 40 ns;
```

The bit operators also operate on bit values, these are **and**, **or**, **nor**, **nand**, **xor** and **not**. Table 11.3 lists the results from these operations. A Boolean and a bit type cannot be mixed in the same operation. For example:

```
bit1<= '0' and False;
```

is illegal because the **and** operator is operating on a bit type and a Boolean type. The context of the logical/bit operators (such as **and** and **or**) is implied by the manner of the operation.

Table 11.3 Bit operations.

A	*B*	*A* **and** *B*	*A* **or** *B*	*A* **nand** *B*	*A* **nor** *B*	*A* **xor** *B*	**not** *A*
0	0	0	0	1	1	0	1
0	1	0	1	1	0	1	1
1	0	0	1	1	0	1	0
1	1	1	1	0	0	0	0

11.12.7 Characters

VHDL uses an 8-bit code to represent characters. The first 127 codes are standard ASCII defined characters, these are listed in Table 11.4.

Table 11.4 ASCII character set.

Hex	*Char*	*Hex*	*Char*	*Hex*	*Char*	*Hex*	*Char*	*Hex*	*Char*	*Hex*	*Char*	*Hex*	*Char*	*Hex*	*Char*
00	NUL	10	DLE	20	SP	30	0	40	@	50	P	60	’	70	p
01	SOH	11	DC1	21	!	31	1	41	A	51	Q	61	a	71	q
02	STX	12	DC2	22	“	32	2	42	B	52	R	62	b	72	r
03	ETX	13	DC3	23	#	33	3	43	C	53	S	63	c	73	s
04	EOT	14	DC4	24	$	34	4	44	D	54	T	64	d	74	t
05	ENQ	15	NAK	25	%	35	5	45	E	55	U	65	e	75	u
06	ACK	16	SYN	26	&	36	6	46	F	56	V	66	f	76	v
07	BEL	17	ETB	27	‘	37	7	47	G	57	W	67	g	77	w
08	BS	18	CAN	28	(	38	8	48	H	58	X	68	h	78	x
09	HT	19	EM	29	)	39	9	49	I	59	Y	69	i	79	y
0A	NL	1A	SUB	2A	*	3A	:	4A	J	5A	Z	6A	j	7A	z
0B	VT	1B	ESC	2B	+	3B	;	4B	K	5B	[	6B	k	7B	{
0C	FF	1C	FSP	2C	,	3C	<	4C	L	5C	\	6C	l	7C	|
0D	CR	1D	GSP	2D	-	3D	=	4D	M	5D	]	6D	m	7D	}
0E	SO	1E	RSP	2E	.	3E	>	4E	N	5E	^	6E	n	7E	~
0F	SI	1F	USP	2F	/	3F	?	4F	O	5F	-	6F	o	7F	DEL

The characters from 0 to 31 (1F, hexadecimal) and character 127 (DEL) are non-printing characters and are enumerated to types, such as NUL, SOH and LF. The characters from 32 to 126 are printable characters and are thus enclosed in quotes, such as ‘A’, Z’ and ‘w’

Examples of uses are:

```
variable in_char,out_char: character;

   in_char=NUL;
   out_char='C';
```

11.12.8 Logic levels

The main objective of VHDL is to model digital hardware. Thus the electrical characteristics of a logic level cannot really be represented by only a '0' and a '1'. For this reason there are other levels. These are:

```
'U'  -- unitialized      'X'  -- forcing unknown
'0'  -- forcing zero     '1'  -- forcing one
'Z'  -- high impedance   'W'  -- weak unknown
'L'  -- weak zero        'H'  -- weak one
'-'  -- don't care
```

The IEEE have standardized the usage of this in a package called `std_logic_1164`. This includes a standard type known as `std_ulogic` which is defined as:

```
type std_ulogic is ('U', 'X', '0', '1',
                    'Z', 'W', 'L', 'H', '-');
```

These levels represent the forcing strength of the signal where it is an output from a tri-state buffers or from pull-up or pull-down loads. The 'U' is an uninitialized state, by default all signals are initialized to this level. Typically 'X' and the '-' state are used when generating test vectors where the level of an input is un-important to the tested part of the system.

The other main forcing lines are 'X', '0' and '1'. When std_ulogic is used then the bit-wise operators (and, or and so on) return either a 'U', 'X', '0' or a '1'. For example if the input to an and gate is an 'X' and a '1' then the output will return 'X' as both inputs to an and gate need to be '1' to give a '1'. The output from a or gate, though, will be '1' because the 'X' input will not have an effect on the output because one of the inputs is a '1'. Table 11.5 shows the output state for a set of inputs for an and-gate. The table gives all the possible combinations. It should be noticed that whenever one of the inputs is a '0' the output is always '0'. Table 11.6 shows the truth table for an or gate. It can be seen that whenever one of the inputs is a '1' the output is always a '1', no matter the state of the other input.

A high-impedance state ('Z') is used to represent the output of a buffer when it is in the high impedance state, or an input which is not connected (and not pulled-up or pulled-down). The weak strength signals ('W', 'L' and 'H') are often used to simulate 'worst-case' situations (or even fault conditions).

Table 11.5 And gate signal inputs and output.

	'U'	'X'	'0'	'1'	'Z'	'W'	'L'	'H'	'-'
'U'	'U'	'U'	'0'	'U'	'U'	'U'	'0'	'U'	'0'
'X'	'U'	'X'	'0'	'X'	'X'	'X'	'0'	'X'	'X'
'0'	'0'	'0'	'0'	'0'	'0'	'0'	'0'	'0'	'0'
'1'	'U'	'X'	'0'	'1'	'X'	'X'	'0'	'1'	'X'
'Z'	'U'	'X'	'0'	'X'	'X'	'X'	'0'	'X'	'X'
'W'	'U'	'X'	'0'	'X'	'X'	'X'	'0'	'X'	'X'
'L'	'U'	'X'	'0'	'1'	'X'	'X'	'0'	'1'	'X'
'H'	'U'	'X'	'0'	'1'	'X'	'X'	'0'	'1'	'X'
'-'	'U'	'X'	'0'	'X'	'X'	'X'	'0'	'X'	'X'

Table 11.6 Or gate signal inputs and output.

	'U'	'X'	'0'	'1'	'Z'	'W'	'L'	'H'	'-'
'U'	'U'	'U'	'U'	'1'	'U'	'U'	'U'	'1'	'U'
'X'	'U'	'X'	'X'	'1'	'X'	'X'	'X'	'1'	'X'
'0'	'U'	'X'	'0'	'1'	'X'	'X'	'0'	'1'	'X'
'1'	'1'	'1'	'1'	'1'	'1'	'1'	'1'	'1'	'1'
'Z'	'U'	'X'	'X'	'1'	'X'	'X'	'X'	'1'	'X'
'W'	'U'	'X'	'X'	'1'	'X'	'X'	'X'	'1'	'X'
'L'	'U'	'X'	'0'	'1'	'X'	'X'	'0'	'1'	'X'
'H'	'1'	'1'	'1'	'1'	'1'	'1'	'1'	'1'	'1'
'-'	'U'	'X'	'X'	'1'	'X'	'X'	'X'	'1'	'X'

The std_ulogic type is not a standard part of the VHDL language but can be included by including the library which defines types, operators and functions for the IEEE standard 1164. To do this the following lines are added near the top of the program:

```
library ieee;
use ieee.std_logic_1164.all;
```

11.13 Flow control

Flow controls allow the VHDL code make decisions (with the if statement) or to branch (with the case statement) or to iterate (with the for statement).

11.13.1 Process control with the process statement

The process statement contains a sensitivity list after the keyword process. This is a list of signals to which the process is sensitive. When any of these signals change value, the process resumes and executes the sequential statements within the process. VHDL block 11.3 shows an example of the process statement. In this case, when any of the signals SELECT, DO, D1 or D2 change then the process mux_pr is automatically called. The standard format for the process statement is:

```
process_name: process(sensitivity_list)
begin
    { sequential_statement(s) }
end process_name;
```

Note that the process name (in this case, it is *process_name*) is optional. Once the process statement is called then the statements within the process are executed sequentially.

11.13.2 Concurrent and sequential signal assignments

Signal assignments within the process statement are executed sequentially (that is, one by one). Assignments outwith a process statement are executed concurrently and are triggered by an event, that is, they are executed whenever there is an event on a signal that appears in its expression. Sequential statements are not event triggered and are executed in a sequence relating to the other sequential statements in the process. An example of concurrent and sequential statements is given next. In the first example (`SEQ_SIGNALS`) the process is called when the `sig1` changes state. When it does then the two statements within the process are executed in one after another. First the `sig2` takes on the value of `sig1` then after this has been executed, `z` takes on the value of `sig2`. There is no inherent time delay between the statements but because one statement is executed before the next there is no requirement for the second statement to take place at some time after the first statement.

In the concurrent version (`CON_SIGNALS`), a change in `sig1` causes the first statement to be executed and thus `sig2` will take the value of `sig1`. Then because `sig2` has changed its state the next time step of the simulator will cause the second statement to be executed and `z` will take on the value of `sig2`.

```
architecture  SEQ_SIGNALS  of  EXAMPLES is
begin
   process(sig1);

   begin
      sig2<=sig1;
      z<=sig2;
   end process;
end architecture;

architecture  CON_SIGNALS  of  EXAMPLES is
begin
   sig2<=sig1;
   z<=sig2;
end architecture;
```

11.13.3 Case statement

The case statement allows a model to branch in a execution to one of more sequential statements. The standard format is:

```
case Boolean_statement is
   when choices => { sequential statement(s) }
   when choices => { sequential statement(s) }
   { ... }

end case;
```

VHDL block 11.3 uses the case statement to implement a 4-bit address multiplexer. It has four data input lines: `D0`, `D1`, `D2` and `D3`, an address line input: `select` and an output `Z`. If the input address line is a zero then the output will be equal to the `D0` input, if it is a one then it will equal the `D1` input, and so on.

As previous mentioned a change on any of the signals in the process sensitivity list causes the `mux_pr` process to be called. In the case of the multiplexer a change on any of the input data signals or the addressing line will cause the output to change its state.

VHDL block 11.3

```
entity  mux is
   port(select,D0,D1,D2,D3: in bit; Z: out bit);
end     mux;

architecture mux_blk  of  mux  is
begin

   mux_pr: process(select,DO,D1,D2,D3) is
   begin
      case SELECT is
         when 0 => Z=D0;
         when 1 => Z=D1;
         when 2 => Z=D2;
         when 3 => Z=D3;
      end case
   end process mux_pr;

end architecture  MUX_BLK;
```

11.13.4 If statement

The if statement allows the various different routes to be taken through a model. It is similar to the if statement used in Pascal and C but does not have a begin (or { in C) and an end (or } in C) to delimit the start and end of conditional statements. The standard format is:

```
if Boolean_statement then
        { sequential statement(s) }
elsif  Boolean_statement then
        { sequential statement(s) }
else
        { sequential statement(s) }
end if;
```

As with the case statement the statements within the `if` statement are sequential statements. VHDL block 11.4 gives a model for an RS flip-flop. In this example if the set is a '1' and reset is a '0' the `q` will become a '1' (after 15 ns) and `qbar` will become a '0' (after 10 ns). The if statement is ended with the `endif` keyword. If else statements are required then the `elsif` keyword is used.

VHDL block 11.4

```
entity  SRFF is
  port(set,reset: in bit; q,qbar: out bit);
  -- truth table S  R  Q  Qbar
  --             0  0  Can't happen
  --             0  1  0    1
  --             1  0  1    0
  --             1  1  1    1
end      SRFF;

architecture SR_BLK  of SRFF is
begin
  sr_proc: process(set,reset) is
  begin
    if set = '1' and reset = '0' then
      q<='0' after 10 ns;
      qbar<='1' after 15 ns;
    elsif set ='0' and reset='1' then
      q<='1' after 15 ns;
      qbar<='0' after 10 ns;
    elsif set ='1' and reset='1' then
      q<='1' after 15 ns;
      qbar<='1' after 15 ns;
    endif
  end process sr_proc;
end architecture  SR_BLK;
```

11.13.5 Wait statement

The wait statement allows a process to be suspended until a given event. This is an alternative to using the process statement which also suspends a process but the wait statement also allows for Boolean and time-based expressions. The basic syntax is:

```
wait on sensitivity list;
wait until Boolean-expression;
wait for time-expression;
```

A few examples are:

```
wait on sig1, sig2, sig3;
wait for 50 ns;
wait until (sig1/=sig2);
```

VHDL block 11.5 is an example of a half adder model which uses the wait statement. In this model the sum and carry are assigned and then the process suspends until there is a change on either a or b. The process is automatically recalled and the sum and carry bits are reassigned. This type of operation is almost identical to using a and b in the sensitivity list. VHDL block 11.6 uses the `wait on` statement to with a Boolean operation wait for `clock` to be set at a '1'.

It is also possible to group expressions in the wait list. For example:

```
wait on clock for 100 ns;    -- statement 1
wait on sig1 until clk='1';  -- statement 2
wait until set='0' for 1 us; -- statement 3
```

Statement 1 waits for the `clock` signal to change within a period of 100 ns. If it does not change within this time then the `wait` statement will exit. In the second statement the wait will wait for `sig1` until the `clk` signal becomes a '1'. If it does become a '1' then the wait statement exits. The third statement waits for set to become a '0' within a period of 1 μs.

VHDL block 11.5

```
entity half_add is
  port(a,b: in bit; sum,carry: out bit);
end entity half_add;

architecture half_blk of half_add is
  half_adder: process is

  begin
    sum<=a xor b;
    carry<=a and b;
    wait on a,b;
  end process half_adder;

end architecture half_add;
```

11.13.6 Loop statement

Iterative loops are often required in a model, for this purpose VHDL uses the `loop` statement to repeat a set of sequential statement until infinity, sometimes called an infinite loop. This type of statement is hardly ever used in high-level programs but is used in VHDL because many electronic systems often stay within a given loop while the power is on. Normally in VHDL the loop statement is used with a `wait` statement to wait for a change in the signal.

VHDL block 11.6 implements a 4-bit counter which starts at a value of 15 and counts down to 0, then is reset back to 15. It uses the `loop` statement to implement an infinite loop. The trigger for the count is the clock signal. Within the infinite loop the program waits at the `wait until clock ='1'` statement until `clock` goes high. After this happens the `count_val` is decremented. Next the `counter_val` variable is tested to determine if it is negative. If it is then it is set to 15. The `counter_val` variable is then assigned to the count and the loop then continues indefinitely.

VHDL block 11.6

```
entity  counter16 is
  port(clock: in bit; count_val: out integer);
end entity counter16;
architecture  count_blk  of counter16 is
begin
  incr_count: process is
  variable  count_val:integer:=15;
  begin
    count<=count_value;
    loop
      wait until clock='1';
      count_val:=(count_val-1);
      if (count_val=-1) then count_val:=15;
      count<=count_val;
    end loop;
  end process incr_count;
end architecture count_blk;
```

11.13.7 For statement

The for statment allows for repetitive looping of a set of sequential statements at specified amount of times.

for *identifier* **in** *discrete_range* **loop**
  { sequential statements }
end loop;

An example of a loop that assigns the values of 0 to 255 into a signal x with a 100 ms delay between each assignment is given next.

```
for i in 0 to 255 loop
   x<=i;
   wait 100 ms;
end loop;
```

It is also possible to count down with the `downto` keyword, as given next.

```
for i in 255 downto 0 loop
   x<=i;
   wait 100 ms;
end loop;
```

11.14 Exercises

11.1 Define data types for the following ranges given:

(a) byte (8-bit signed integer);
(b) unsigned_byte (16-bit unsigned integer);
(c) short_int (16-bit signed integer);
(d) unsigned_short_int (16-bit unsigned integer);
(e) long_int (32-bit signed integer);
(f) unsigned_long_int (32-bit unsigned integer).

11.2 With reference to Table 11.5 and Table 11.6, and with the `std_ulogic` types, complete the truth table in Table 11.7 for a not gate.

Table 11.7 Not gate truth table.

Input	'U'	'X'	'0'	'1'	'Z'	'W'	'L'	'H'	'-'
Output									

11.3 With reference to Table 11.5 and Table 11.6 determine a truth table, with the `std_ulogic` types, complete Table 11.8, for each of the following:

(a) Exclusive-OR gate;
(b) Nor gate;
(c) Nand gate.

11.4 VHDL block 11.3 allows the multiplexer select line `select` to become any value. Rewrite the block with a type definition so that the value of `select` can only range between 0 and 3.

Table 11.8 Logic gate inputs and output.

	'U'	'X'	'0'	'1'	'Z'	'W'	'L'	'H'	'-'
'U'									
'X'									
'0'									
'1'									
'Z'									
'W'									
'L'									
'H'									
'-'									

11.5 Discuss the differences between concurrent statements and sequential statements.

11.6 Write a VHDL model for a D-type flip-flop (which is level-triggered). The basic specification is:

if the CLEAR input is a '1' then the output Q is a '0';
else, if the CLOCK is a '1' then the Q output becomes the D input.

PSpice Modelling

12

12.1 Introduction

Most of this book contains digital electronic applications. In many cases it is only necessary to simulate and design with digital logic. Unfortunately, these digital device are not perfect. Each has its own voltage level, its own low-to-high risetime, its own high-to-low risetime, its own input and output impedance, its own drive-current, and so on. To model these parameters an analogue simulation is required. One of the most used models is SPICE. It was originally developed by the University of California and is now available on many different computer systems and electronic design packages.

This chapter uses the version of SPICE known as PSpice. The software used is the evaluation copy of the MicroSim Corporation PSpice for Windows, which is available over the Internet (see Appendix H). A tutorial example using PSpice in given in Appendix F.

12.2 DC analysis

Spice uses a node-based approach to describe the layout of circuit. Figure 12.1 shows a simple resistor-based circuit with a 12 V DC supply. It can be seen that the equivalent resistance of the R_2 resistor in parallel with R_3 and R_4 will be 1 kΩ. The voltage between node 2 and 0 will thus be 6 V and the input current from the voltage source will be 6 mA because the input resistance is 2 kΩ. The Spice model for this circuit is given in PSpice block 12.1.

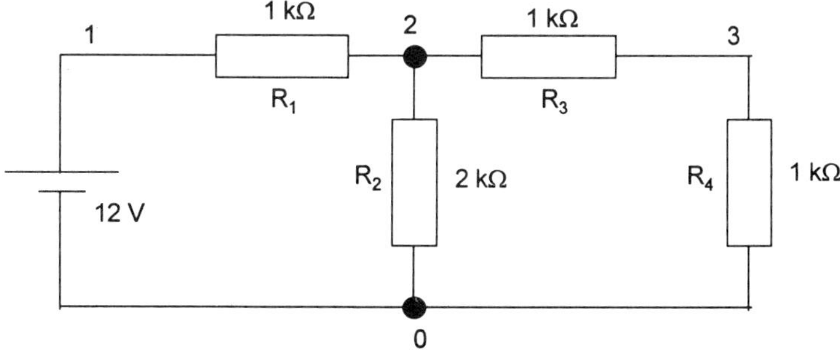

Figure 12.1 Simple resistor circuit.

Extra instructions to the simulator are added with dot commands (*.cmd*). The `.OP` instruction gives operating point instructions on the nodes in the simulation. In this case, the node voltages are 12V, 6V and 3V (on nodes 1, 2 and 3, respectively). The first line of PSpice file is always taken as a comment.

PSpice block 12.1

```
Example program
VS 1 0 12V
R1 1 2 1k
R2 2 0 2k
R3 2 3 1k
R4 3 0 1k
.op
.end
```

Test run 12.1 shows the resulting output from the simulation. It shows that the voltages at nodes 1, 2 and 3 are 12 V, 6 V and 3 V, respectively. The simulation also shows that the input current is 6 mA (as expected) and the total power dissipation is 72.0 mW ($12^2/2000$).

PSpice test run 12.1

```
****SMALL SIGNAL BIAS SOLUTION    TEMPERATURE =27.000 DEG C
NODE VOLTAGE     NODE     VOLTAGE NODE     VOLTAGE
(1)   12.0000    (2)      6.0000   (3)     3.0000
    VOLTAGE SOURCE CURRENTS
    NAME            CURRENT
    vs             -6.000E-03
    TOTAL POWER DISSIPATION    7.20E-02   WATTS
```

A DC analysis is conducted using the statement `.DC VS 12V 12V 0V`. This gives a DC analysis with VS starting at 12 V, ending at 12 V with a step size of 0 V. The current within the resistors can be printed using the `.PRINT DC I(R1) I(R2) I(R3)`. Test run 12.2 shows that these currents are 6 mA, 3 mA and 3 mA (as expected). The standard format for the .DC statement is:

`.DC` *variable start end increment*

PSpice block 12.2

```
Example program
VS 1 0 12V
R1 1 2 1k
R2 2 0 2k
R3 2 3 1k
R4 3 0 1k
.OP
.PRINT DC I(R1) I(R2) I(R3)
.dc vs 12V 12V 0V
.end
```

PSpice test run 12.2

```
****DC TRANSFER CURVES          TEMPERATURE =    27.000 DEG C

  VS            I(R1)         I(R2)         I(R3)
   1.200E+01     6.000E-03     3.000E-03     3.000E-03


****SMALL SIGNAL BIAS SOLUTION    TEMPERATURE =27.000 DEG C

NODE  VOLTAGE     NODE     VOLTAGE  NODE     VOLTAGE
(1)   12.0000     (2)      6.0000   (3)      3.0000

    VOLTAGE SOURCE CURRENTS
    NAME          CURRENT
    VS           -6.000E-03

    TOTAL POWER DISSIPATION    7.20E-02   WATTS
```

12.3 PSpice simulation

The PSpice schematic of the circuit in Figure 12.1 is shown in Figure 12.2. The netlist of the PSpice parameters produced is given in PSpice block 12.3. Appendix F contains more information on how to enter schematics in PSpice and how to simulate them. Notice that, in this case, the schematic wire connections have not been given labels and thus the netlist has assigned default labels, such as $N_0001, $N_0002, and so on.

PSpice block 12.3

```
* C:\MSIMEV60\NEW.SCH
.OP

R1          R1(1=$N_0002 2=$N_0001 )
R2          R2(1=$N_0001 2=$N_0003 )
R3          R3(1=0 2=$N_0001 )
R4          R4(1=0 2=$N_0003 )
V1          V1(+=$N_0002 -=0 )

.probe
.end
```

12.4 Units

Spice allows for values to be in exponent format (such as 1.4e–10) or they can be specified in a symbolic form. Table 12.1 lists the main unit symbols and examples of typical units are given in Table 12.2.

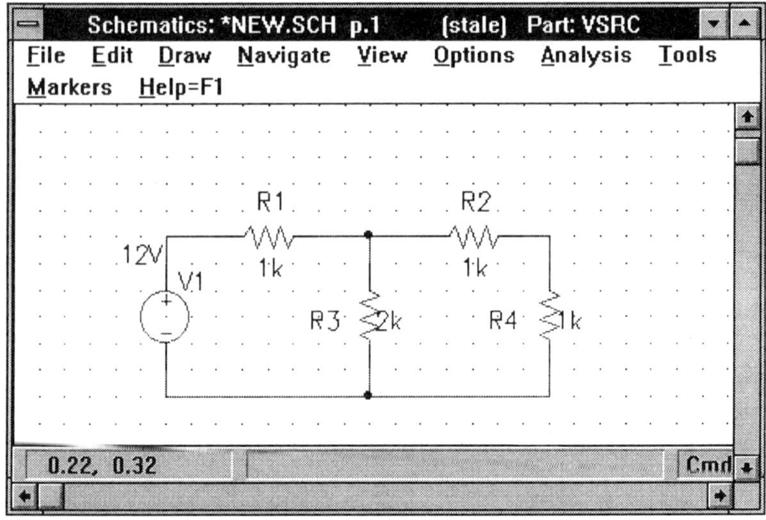

Figure 12.2 PSpice resistor circuit.

Table 12.1 Symbolic forms for units.

Value	*Symbolic form*	*Exponential form*
10^{-15}	f	1e–15
10^{-12}	p	1e–12
10^{-9}	n	1e–9
10^{-6}	u	1e–6
10^{-3}	m	1e–3
10^{3}	k	1e3
10^{6}	meg	1e6
10^{9}	g	1e9
10^{12}	t	1e12

Table 12.2 Symbolic forms for units.

Value	*Symbolic form*
Capacitance	pF, nF, uF, mF
Resistance	k, meg
Inductance	nH, uH, mH, H
Time	ps, ns, us, ms, s
Frequency	Hz, kHz, megHz
Current	uA, mA, A
Voltage	nV, uV, mV, V

12.5 Device references and model parameters

Spice is not case sensitiive and all text can be written in either upper or lower case. Comment are inserted after a * or a semicolon (;). The type of component in a Spice model is identified by the first letter of its name. The main device references and model parameters:

- B GaAsFet
- C Capacitor
- D Diode
- E Voltage-Controlled Voltage Source
- F Current-Controlled Current Source
- G Voltage_Controlled Current Source
- H Current-Controlled Voltage Source
- I Independent Current Source and Stimulus
- J Junction FET
- K Inductor-Coupling (transformer core)
- L Inductor
- M MOSFET
- N Digital Input
- O Digital Output
- Q Bipolar Transistor
- R Resistor
- S Voltage-Controlled Switch
- T Transmission Line
- U Stimuli and Primitives
- V Independent Voltage Source and Stimulus
- W Current-Controlled Switch
- X Subcircuit Call

Table 12.3 outlines the main electrical devices and their summarized format (Their full format is given in Appendix G). Thus a component named Q12 is a bi-polar transistor, R111 is a resistor, C1 is a capacitor and D2 is a diode. The format descriptors in brackets [*name*] are optional and descriptors within angled brackets, that is, <*name*>, specify required items. Example of components are:

```
Q12 4 5 6 Q2N2222A      ; bi-polar transistor connected to
                        ; ports 4, 5 and 6 which uses model
                        ; Q2N2222A

L 1 2 10mH              ; a 10 mH inductor connected
                        ; between nodes 1 and 2

C 3 0 100uF             ; a 100 uF capacitor connected
                        ; between nodes 3 and 0
```

Table 12.3 Outline format for components.

Component	*Format*
GaAs FET	B[*name*] <*drain*> <*gate*> <*source*> <*model_name*>
Capacitor	C[*name*] <*+node*> <*-node*> [*model_name*] <*value*>
Diode	D[*name*] <*+node*> <*-node*> <*model_name*>
JFET	J[*name*] <*drain*> <*gate*> <*source*> <*model_name*>
Inductor	L[*name*] <*+node*> <*-node*> <*value*>
MOSFET	M[*name*] <*drain*> <*gate*> <*source*> <*bulk_node*> <*model_name*>
Bi-polar trans.	Q[*name*] <*collector*> <*base*> <*emitter*> <*model_name*>
Trans. Line	T[*name*] <*+Anode*> <*-Anode*> <*+Bnode*> <*-Bnode*> <*Zo*> <*td*> <*f*> <*nl*>

12.6 Voltage sources and stimulus

An independent voltage source is often used as the stimulus to a circuit and is identified by a preceding V (such as V1, V10, and so on). The first two parameters after its reference name are the node which it connects to. In general, the format is:

V<*name*> <*+node*> <*–node*> [[*DC*] <*value*>]
\+ [AC <*magnitude_value*> [*phase_value*]]
\+ [*transient_specification*]

The + symbol joins of multiple lines into a single statement line and the brackets ([*option*]) specify optional parameters. Default values for the DC, AC, and transient values are zero. The statement can contain all, or none, of the DC, AC, and transient values. An AC phase value is specified in degrees. Table 12.4 gives examples of Spice statements for voltage sources.

Table 12.4 Example statements for voltage sources.

Spice statement	*Description*
`vin 13 0 10mV`	DC voltage of 10 mV between nodes 13 and 0.
`vsource 2 3 AC 0.001`	AC source of 0.001 volts.
`Vs 1 2 ac 1 90`	AC voltage source of 1 V at a phase angle of 90°.
`vpulse 1 0 pulse(-1V 1V 1ms 2ns 2ns 10ms 30ms)`	Voltage pulse of initial amplitude of –1V, end voltage of 1 V, delay time of 1 ms, rise time of 2 ns, fall time of 2 ns, pulse width of 10 ms and period of 30 ms.

Table 12.4 (Cont.).

Spice statement	*Description*
`vin 10 5 exp(2V 10V 1ms 0.1ms 2ms 0.1ms)`	Exponential voltage with an initial voltage of 2 V, a peak voltage of 10 V, a rise delay of 1 ms, a rise time constant of 0.1 ms, a fall delay of 2 ms and a fall time constant of 0.1 ms.
`v1 3 4 DC 10mV ac 1 sin(10mV 100mV 1meg)`	AC voltage source with a DC voltage of 10 mV, an amplitude of 100 mV at a frequency of 1 MHz.
`V3 3 4 pwl(0 0V 0.5 2V 1 5V 1.5 2.3V 2.1 5V 3 -1V)`	Piecewise waveform which starts at 0 V at 0 s, rises to 2 V at 0.5 s, rises to 5 V at 1 s, falls to 2.3 V at 1.5 s and so on.
`vfm 1 5 sffm(2V 1V 10kHz 2 0.5kHz)`	Single frequency FM waveform with an offset voltage of 2 V, a peak voltage of 1 V, a carrier frequency of 10 kHz, a modulation index of 2 and a modulation frequency of 0.5 kHz.

Voltage sources can also contain a transient specification, which is one of the following:

```
EXP(<parameters>)    ; an exponential waveform
PULSE(<parameters>)  ; a pulse waveform
PWL(<parameters>)    ; a piecewise linear waveform
SFFM(<parameters>)   ; a frequency-modulated waveform
SIN(<parameters>)    ; a sinusoidal waveform
```

Transient parameters can be specified with the `.TRAN` command. The format of this is:

```
.TRAN <tstep_value> <tfinal_value>
```

These times can also be set within the transient waveform specifications. The `.TRAN` command is normally used when the transient times values are common to the complete simulation.

12.6.1 Exponential transient

The general format for the exponential waveform is:

```
EXP(<v1> <v2> <td1> <τ1> <td2> <τ2>)
```

where *<v1>* – initial voltage (volt). Default: none.

<*v2*> – peak voltage (volt). Default: none.
<*td1*> – rise delay (sec). Default: 0.
<*τ1*> – rise time constant (sec). Default: *tstep_value*
<*td2*> – fall delay (sec). Default: 0.
<*τ2*> – fall time constant (sec). Default: *tstep_value*.

12.6.2 Pulse transient

The general form for a pulse waveform is

PULSE (<*v1*> <*v2*> <*td*> <*tr*> <*tf*> <*pw*> <*per*>)

where <*v1*> – initial voltage (volt). Default: none.
<*v2*> – pulsed voltage(volt). Default: none.
<*td*> – delay (sec). Default: 0.
<*tr*> – rise time (sec). Default: *tstep_value*.
<*tf*> – fall time(sec). Default: *tstep_value*.
<*pw*> – pulse width (sec). Default: *tstop_value*.
<*per*> – period (sec). Default: *tstop_value*.

12.6.3 Piecewise linear transient

The general form of a piecewise linear waveform is:

PWL (<*t1*> <*v1*> <*t2*> <*v2*> ... <*tn*> <*vn*>)

where <*tn*> – time for voltage change (sec). Default: none.
<*vn*> – voltage level at *tn* (volt). Default: none.

12.6.4 Single-frequency FM transient

The general form of the single-frequency FM waveform is:

SFFM (<*voff*> <*vamp*> <*fc*> <*mod*> <*fm*>)

where <*voff*> – offset voltage (volt). Default: none.
<*vamp*> – peak amplitude of voltage (Volt). Default: none.
<*fc*> – carrier frequency (Hz). Default: 1/ *tstop_value*.
<*mod*> – modulation index. Default: 0.
<*fm*> – modulation frequency (Hz). Default: 1/ *tstop_value*.

12.6.5 Sine wave transient

The general form for a sine wave is:

SIN (<voff> <vampl> <freq> <td> <df> <phase>)

where *<voff>* – offset voltage (volt). Default: none.
<vampl> – peak amplitude of voltage (volt). Default: none.
<freq> – frequency (Hz). Default: 1/ *tstop_value*.
<td> – delay (sec). Default: 0.
<df> – damping factor (1/s). Default: 0.
<phase> – phase (degree). Default: 0.

The `SIN` transient starts at a voltage of *<voff>* and stays there for *<td>* seconds. After this the voltage becomes an exponentially damped sine wave. This waveform should only be used for transient analysis only and it has no effect during AC analysis.

12.7 Current sources and stimulus

An independent current source is less often used as the stimulus than a voltage source but is typical used when modelling active devices. It is identified by a preceding `I` (such as `I1`, `Iramp`, and so on). The first two parameters after its reference name are the node which it connects to. In general, the format is:

I*<name>* *<+node>* *<–node>* [[*DC*] *<value>*]
\+ [AC *<magnitude_value>* [*phase_value*]]
\+ [*transient_specification*]

This format is similar to the voltage source definition in Section 12.6. Example are:

```
Iinput 5 0 1.7mA
Iac 2 3 AC 1mV
Iin 1 3 AC 1mV 90
Ipulse_in 1 0 pulse(-10mA 10mA 2us 2us 2us 50us 80us)
I3 3 4 DC 2mV AC 1 SIN(2mV 2mV 0.002 1.5MEG)
```

12.8 Dot commands

The dot commands provide the simulator with the extra information required to conduct the analysis. Table 12.5 lists these.

Table 12.5 Dot commands.

Dot command	*Description*
`.AC` *<points>* *<fstart>* *<fend>*	AC sweep which starts at *fstart* and simulates up to *fend*. The number of frequency points in-between is points

Table 12.5 (Cont.).

Dot command	*Description*
`.DC` <*sweep_variable*> <*start*> <*end*> <*increment*>	DC analysis for *sweep_variable* from *start* to *end* with a voltage increment of *increment*
`.END`	End of file
`.FOUR` <*frequencies*> <*output variable*>	Conduct Fourier analysis for the specified frequencies
`.INC` <*filename*>	Insert another file
`.LIB` <*file_name*>	References a model in another library (in PSpice this is EVAL.LIB)
`.MODEL` [*name*] [*type*]	Describes a model
`.NOISE` <*Vnode*> [,*node*] <*name*>	Gives noise analysis along with AC analysis
`.OP`	Gives detailed bias-point information
`.OPTIONS` <*option_name*>	Sets options for analysis
`.PLOT` [*DC*] [*AC*] [*NOISE*] [*TRAN*] [*output_variable*]	Gives printer-type plot
`.PROBE` [*output_variable*]	Creates PROBE.DAT file for graphics analysis
`.SENS` <*output_variable*>	Sensitivity analysis
`.TEMP` <*temperature*>	Sets temperature of analysis
`.TF` <*output_variable*> <*input_variable*>	Transfer function, which is the ratio of the *output_variable* to the *input_variable*

12.8.1 Examples

PSpice block 12.4 is a series RC circuit with a resistance of 1 kΩ and a capacitance of 1 nF. The `.AC` command has been used to simulate the model from 10 Hz up to 1 MHz for 20 linear frequency steps. The `.print` command is then used to print a table of the result of an ac simulation with the voltage across R1 (which is specified by v(r1)) and the voltage across C1 (which is specified by v(c1)). Test run 12.3 shows the test results from a PSpice simulation. Notice that as the frequency increases the voltage across the capacitor decreases (because its impedance reduces). The circuit thus acts as a low-pass filter.

PSpice block 12.4

```
V1      1 0   AC 1V
R1      1 2 1k
C1      0 2 1n
.ac lin 20 10 1meghz
.print ac v(r1) v(c1)
.end
```

PSpice test run 12.3

```
    FREQ          V(r1)       V(c1)
  1.000E+01   6.283E-05    1.000E+00
  5.264E+04   3.140E-01    9.494E-01
  1.053E+05   5.517E-01    8.341E-01
  1.579E+05   7.043E-01    7.099E-01
  2.105E+05   7.977E-01    6.030E-01
  2.632E+05   8.557E-01    5.175E-01
  3.158E+05   8.930E-01    4.501E-01
  3.684E+05   9.180E-01    3.966E-01
  4.211E+05   9.354E-01    3.536E-01
  4.737E+05   9.479E-01    3.185E-01
  5.263E+05   9.572E-01    2.894E-01
  5.790E+05   9.642E-01    2.651E-01
  6.316E+05   9.697E-01    2.444E-01
  6.842E+05   9.740E-01    2.266E-01
  7.368E+05   9.775E-01    2.111E-01
  7.895E+05   9.803E-01    1.976E-01
  8.421E+05   9.826E-01    1.857E-01
  8.947E+05   9.845E-01    1.751E-01
  9.474E+05   9.862E-01    1.657E-01
  1.000E+06   9.876E-01    1.572E-01
```

PSpice block 12.5 again uses the `.print` command to output a table for the ac response of v(r1) and v(c1). The `.plot` command is used to produce a text-based plot of the V(2) signal, and given in test run 12.4. A graphical plot (produced by PROBE) of this is given in Figure 12.3.

PSpice block 12.5

```
; Low pass filter circuit

V1      1 0   AC 1V
R1      1 2 1k
C1      0 2 1n

.ac lin 20 10 1meghz
.print ac v(r1) v(c1)

.plot ac v(r1) v(c1)
.end
```

PSpice test run 12.4

```
LEGEND:
*: V(r1)
+: V(c1)

  FREQ         V(r1)

(*)----------     1.0000E-05    1.0000E-03    1.0000E-01    1.0000E+01
1.0000E+03
(+)----------     1.0000E-01    1.0000E+00    1.0000E+01    1.0000E+02
1.0000E+03
                            - - - - - - - - - - - - - - - - - - - -
  1.000E+01   6.283E-05 .       *        .+           .            .
  5.264E+04   3.140E-01 .                +            .    *       .
  1.053E+05   5.517E-01 .               +.            .     *      .
  1.579E+05   7.043E-01 .               +.            .      *     .
  2.105E+05   7.977E-01 .              + .            .      *     .
  2.632E+05   8.557E-01 .             +  .            .       *    .
  3.158E+05   8.930E-01 .            +   .            .       *    .
  3.684E+05   9.180E-01 .           +    .            .       *    .
  4.211E+05   9.354E-01 .           +    .            .       *    .
  4.737E+05   9.479E-01 .          +     .            .       *    .
  5.263E+05   9.572E-01 .          +     .            .       *    .
  5.790E+05   9.642E-01 .         +      .            .       *    .
  6.316E+05   9.697E-01 .         +      .            .       *    .
  6.842E+05   9.740E-01 .        +       .            .       *    .
  7.368E+05   9.775E-01 .        +       .            .       *    .
  7.895E+05   9.803E-01 .       +        .            .       *    .
  8.421E+05   9.826E-01 .       +        .            .       *    .
  8.947E+05   9.845E-01 .       +        .            .       *    .
  9.474E+05   9.862E-01 .      +         .            .       *    .
  1.000E+06   9.876E-01 .      +         .            .       *    .
                            - - - - - - - - - - - - - - - - - - - -
```

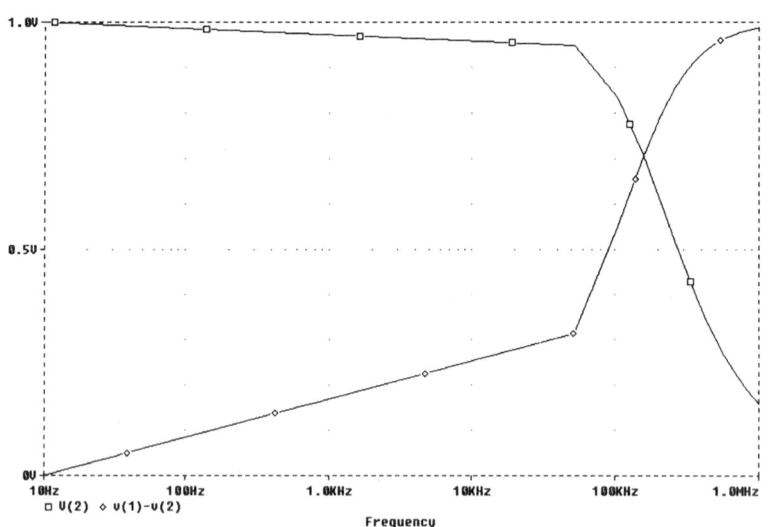

Figure 12.3 Voltage plot.

12.9 Active components

The parameters required for active components, such as bi-polar transistors and FET transistors are normally specified in data sheets or are contained in a library of components.

12.9.1 MOSFET

The general form of a MOSFET transistor is:

```
M<name> <drain node> <gate node> <source node>
+   <bulk/substrate node> <model name>
+   [L=<value>] [W=<value>]
+   [AD=<value>] [AS=<value>]
+   [PD=<value>] [PS=<value>]
+   [NRD=<value>] [NRS=<value>]
+   [NRG=<value>] [NRB=<value>]
+   [M=<value>]
```

The modelling of the device is achieved by setting the physical dimensions of the MOSFET. The actual process parameters, such as absolute resistivity values are specified within the model. The effective length and width of the channel are specified by L and W, respectively (note that is it also possible to specify them in the `.OPTIONS` statement or in a model). If the length and width are not specified then the default, in PSpice, is set at 100 μm.

The source and diffusion areas are specified by AD and AS, respectively and the drain and source diffusion parameters by PD and PS, respectively. Drain, sources, gates and substrate relative resistivity are specified by NRD, NRS, NRG, and NRB, respectively, and are specified in squares. Actual resistivity values are specified within the model.

PD and PS default to 0, NRD and NRS default to 1, and NRG and NRB default to 0. Defaults for L, W, AD, and AS may be set in the `.OPTIONS` statement. If AD or AS defaults are not set, they also default to 0.

M is a device "multiplier" (default = 1), which simulates the effect of multiple devices in parallel. The effective width, overlap and junction capacitances, and junction currents of the MOSFET are multiplied by M, while the parasitic resistance values (such as, RD and RS) are divided by M.

Some example are:

```
M10 1 2 3 0 PNOM  L=25u W=12u
M3 15 3 0 0 PSTRONG
M6 1 3 0 0 PSTRONG M=2
M12 0 2 6 6 NWEAK L=33u W=12u
+  AD=288p AS=288p PD=60u PS=60u NRD=14 NRS=24 NRG=10
```

The models (in the case of the examples are named PMON, PSTRONG, NWEAR) are defined in the form:

```
.MODEL <model name> NMOS [model_parameters]
.MODEL <model name> PMOS [model_parameters]
```

The MOSFET is modelled as an intrinsic MOSFET with ohmic resistances in series with the drain, source, gate, and bulk (substrate). There is also a shunt resistance (R_{DS}) in parallel with the drain-source channel.

12.9.2 Bi-polar transistor

The general form of a bi-polar transistor is:

```
Q<name> <collector node> <base node> <emitter node>
+    [substrate node] <model name> [area value]
```

Examples are:

```
Q1 14 2 13 PNPNOM

Q13 15 3 0 1 NPNSTRONG 1.5

Q7 VC 5 12 [SUB] LATPNP
```

The model definition is in the form:

```
.MODEL <model name> NPN [model parameters]

.MODEL <model name> PNP [model parameters]

.MODEL <model name> LPNP [model parameters]
```

Refer to Appendix G for more information on the modelling parameters for a bi-polar transistor.

12.10 Example circuits

This section contains a few examples of PSpice simulations. PSpice allows many different types of analysis to be conduction on a simulation. The MicroSim schematic capture package (PSchEd) has a PSpice menu option to be generated from the schematic entry. In PSchEd the form for setting up the analysis type is selected from the Analysis→Setup menu option. Figure 12.4 shows the different analysis types and Figure 12.5 show an example of the AC sweep and noise analysis form. In this case the analysis will be a linear AC sweep from 1 kHz to 10 kHz with 10 steps in-between. The noise option is not enabled.

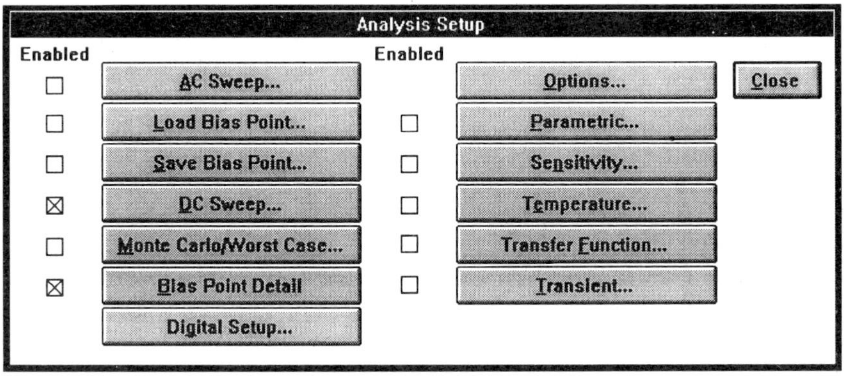

Figure 12.4 Analyis setup in PSpice.

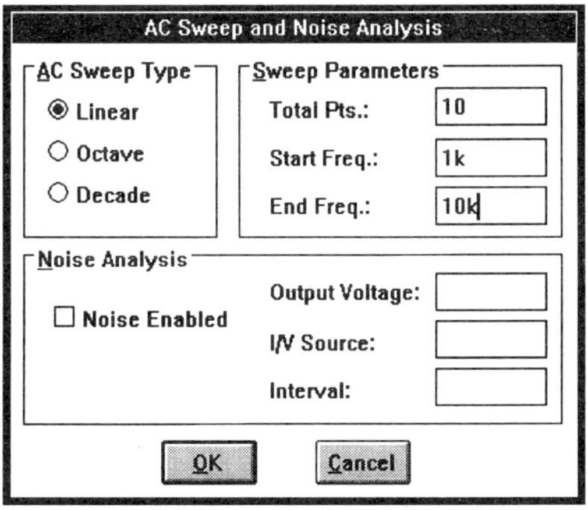

Figure 12.5 Example of the AC sweep and noise analysis form.

12.10.1 Differential amplifier

Figure 12.6 shows a differential amplifier produced by the MicroSim schematic capture package. The filename of the circuit is `DIFF.SCH` and the bi-polar transistor used is the Q2N2222, which is found in the `EVAL.SLB` symbol library. The model for the Q2N2222 is stored in the `EVAL.LIB` file. An excerpt from this file shows the definition of the model:

```
.model Q2N2222  NPN(Is=14.34f Xti=3 Eg=1.11
+ Vaf=74.03 Bf=255.9 Ne=1.307 Ise=14.34f
+ Ikf=.2847 Xtb=1.5 Br=6.092 Nc=2 Isc=0
+ Ikr=0 Rc=1 Cjc=7.306p Mjc=.3416 Vjc=.75
+ Fc=.5 Cje=22.01p Mje=.377 Vje=.75
+  Tr=46.91n Tf=411.1p Itf=.6 Vtf=1.7 Xtf=3 Rb=10)
```

Generating the netlist

The PSpice file produced by PSchEd is given in PSpice block 12.6. As there is no labelling on the nets on the schematic then the nets in the netlist have been assigned names in the form `$N_XXXX`. For example the net connecting the 12 V supply has been given the name `$N_0007`, the net connecting the –6V DC supply is assigned to `$N_0006`, and so on.

PSpice block 12.6

```
* C:\MSIMEV60\DIFF.SCH
* Schematics Version 6.0 - January 1994
* Thu Aug 22 14:36:47 1996
.OP
.lib eval.lib
Q1      $N_0002 $N_0001 $N_0003 Q2N2222
Q2      $N_0005 $N_0004 $N_0003 Q2N2222
R1      $N_0006 $N_0003 5k
R2      $N_0002 $N_0007 1k
R3      $N_0005 $N_0007 1k
R4      $N_0008 $N_0001 1k
R5      $N_0009 $N_0004 1k
V1      $N_0007 0 DC 12V
V2      $N_0006 0 DC -6V
V4      $N_0009 0  AC 0.5mV
V3      $N_0008 0  AC 0.5mV
.END
```

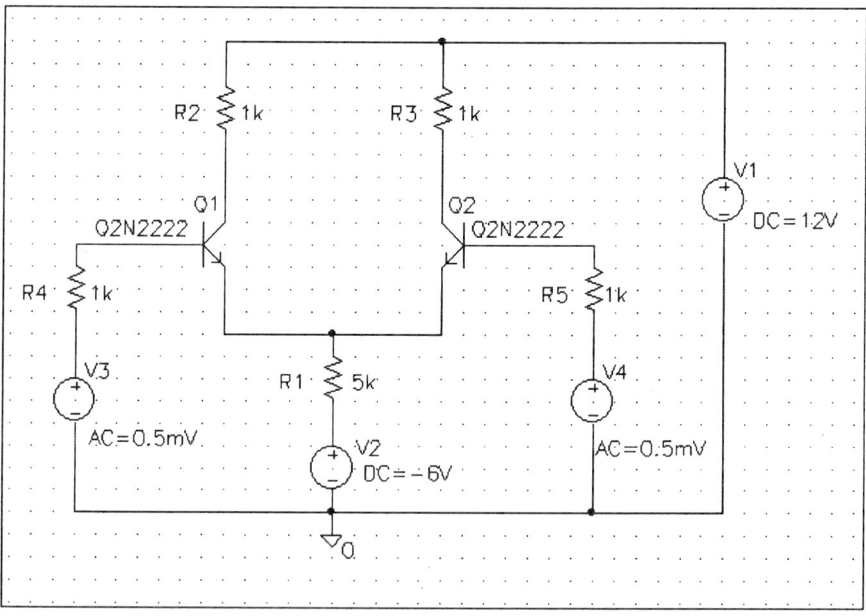

Figure 12.6 Differential amplifier schematic.

The default netlisting for node names is difficult to identify thus a better method is to assign labels to each of the nets; this is shown in Figure 12.7. PSpice block 12.7 shows the new PSpice model.

PSpice block 12.7

```
* C:\MSIMEV60\DIFF.SCH
* Schematics Version 6.0 - January 1994
* Thu Aug 22 14:36:47 1996
.OP

.lib eval.lib
Q1      9 7 4 Q2N2222
Q2      2 3 4 Q2N2222
R1      6 4 5k
R2      9 1 1k
R3      2 1 1k
R4      8 7 1k
R5      5 3 1k
V1      1 0 DC 12V
V2      6 0 DC -6V
V4      5 0  AC 0.5mV
V3      8 0  AC 0.5mV
.END
```

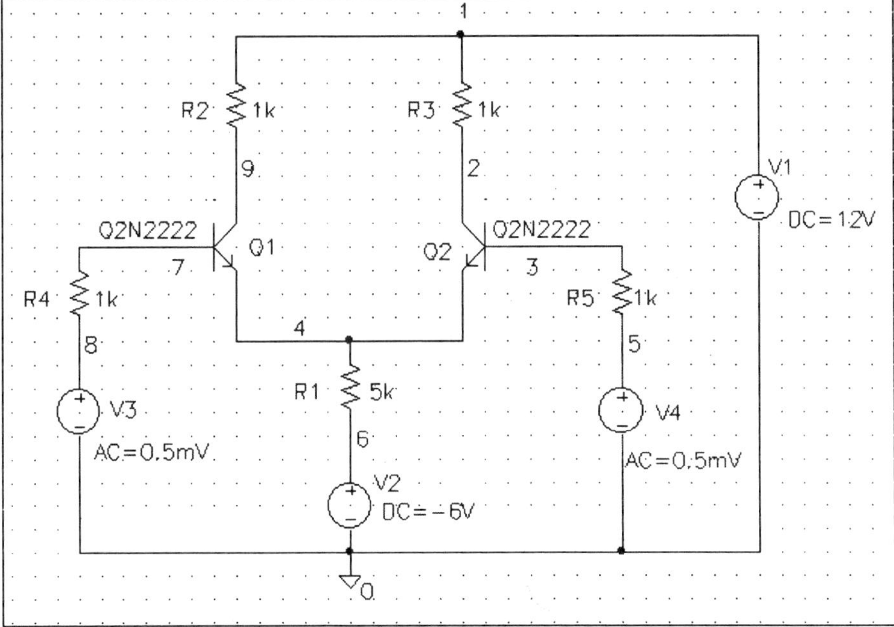

Figure 12.7 Differential amplifier schematic with labels on nets.

DC simulation

PSpice test run 12.5 shows the output file from a DC simulation of the circuit. It can be seen that the voltage on the collector of Q1 and Q2 is 11.466 V, the voltage on the emitters is –0.6293 V and the base voltage is 0.0038 V (due to the base current). Test run 12.5 also shows that the current from the V1 supply is 1.067 mA, from the V2 supply it is 1.074 mA. In can also be seen that there is a DC current of 3.8383 μA flowing through the AC sources.

The bi-polar junction characteristics shows that the collector current is 0.534 mA and the base current is 3.38 μA. A negative value of current indicates that the current is flowing out of the voltage source and a positive value shows a flow of current into the source. The simulation also shows that the base-emitter voltage is 0.626 V and the collector-emitter voltage (V_{CE}) is 12.1 V.

PSpice test run 12.5

```
NODE   VOLTAGE   NODE   VOLTAGE    NODE   VOLTAGE   NODE  VOLTAGE
(1)    12.0000   (2)    11.4660   (3)     -.0034   (4)      -.6293
(5)     0.0000   (6)    -6.0000   (7)     -.0034   (8)      0.0000
(9)    11.4660


    VOLTAGE SOURCE CURRENTS
    NAME           CURRENT
    V1          -1.067E-03
    V2           1.074E-03
    V4          -3.383E-06
    V3          -3.383E-06
    TOTAL POWER DISSIPATION   1.93E-02  WATTS


**** BIPOLAR JUNCTION TRANSISTORS
NAME             Q1          Q2
MODEL            Q2N2222      Q2N2222
IB               3.38E-06     3.38E-06
IC               5.34E-04     5.34E-04
VBE              6.26E-01     6.26E-01
VBC             -1.15E+01    -1.15E+01
VCE              1.21E+01     1.21E+01
BETADC           1.58E+02     1.58E+02
GM               2.06E-02     2.06E-02
RPI              8.58E+03     8.58E+03
RX               1.00E+01     1.00E+01
RO               1.60E+05     1.60E+05
CBE              4.43E-11     4.43E-11
CBC              2.82E-12     2.82E-12
CBX              0.00E+00     0.00E+00
CJS              0.00E+00     0.00E+00
BETAAC           1.77E+02     1.77E+02
FT               6.96E+07     6.96E+07
```

Transient analysis

A transient analysis can be conducted on the circuit by adding the `.TRAN` option. In the PSpice block 12.8 the transient analysis is conducted from 0.1 ms to 2 ms. The transient source added to V4 is a sine wave with no DC offset, an ac amplitude of 2 mV and a frequency of 1000 Hz, with no time delay, zero damping factor, and no phase delay (`sin(0 10mV 1000 0 0 0)`). The ac transient wave applied to the V3 supply has a zero amplitude.

Figure 12.8 shows a plot of the voltage on V(9) and V(2). Notice that as V3 increases the voltage V(9) decreases. This is because the current in Q1 increases causing an increase in the voltage dropped across R2. At the same time the voltage at V(2) increases because there is an equivalent reduction in the current flowing through Q2.

PSpice block 12.8

```
* Schematics Version 6.0 - January 1994
* Thu Aug 22 14:36:47 1996

.OP
.lib eval.lib
Q1      9 7 4 Q2N2222
Q2      2 3 4 Q2N2222

R1      6 4 5k
R2      9 1 1k
R3      2 1 1k
R4      8 7 1k
R5      5 3 1k

V1      1 0 DC 12V
V2      6 0 DC -6V
V4      5 0 sin(0 10mV 1000 0 0 0)
V3      8 0 sin(0 0 1000 0 0 0)

.tran 0.1ms 2ms
.probe
.end
```

AC simulation

An AC simulation is conducted with `.AC` option and a transfer response can be generated with the `.TF` option. PSpice block 12.9 contains these options and test run 12.6 shows the results of the simulation.

In this case it can be seen that at 1 kHz that the ratio of V(9) to the input voltage source is –9.2. The input resistance at V3 is 19.08 kΩ and the output resistance at V(9) is 996.9 Ω.

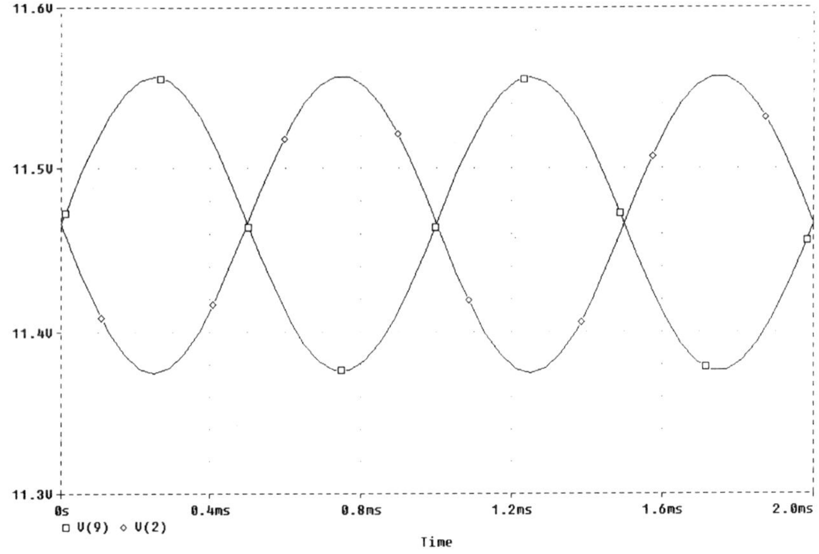

Figure 12.8 Voltage output on V(2) and V(9).

PSpice block 12.9

```
* C:\MSIMEV60\DIFF.SCH
.OP
.lib eval.lib
Q1      9 7 4 Q2N2222
Q2      2 3 4 Q2N2222
R1      6 4 5k
R2      9 1 1k
R3      2 1 1k
R4      8 7 1k
R5      5 3 1k
V1      1 0 DC 12V
V2      6 0 DC -6V
V4      5 0  AC 0.5mV
V3      8 0  AC 0.5mV
.ac LIN 1 1kHz 1KHz
.TF v([9]) V3
```

PSpice test run 12.6

```
****      SMALL-SIGNAL CHARACTERISTICS

      V(9)/V3 = -9.206E+00

      INPUT RESISTANCE AT V3 =  1.908E+04
      OUTPUT RESISTANCE AT V(9) =  9.969E+02
```

DC transfer characteristics

The DC transfer characteristics of a circuit can be generated using the `.DC` option. PSpice block 12.10 shows the differential amplifier circuit with a simulation of its DC characteristics from an input on V3 of –1 V to 1 V in increments of 0.1 V. Figure 12.9 shows a plot of the results (using PROBE).

PSpice block 12.10

```
* C:\MSIMEV60\DIFF.SCH
.OP
.lib eval.lib
Q1      9 7 4 Q2N2222
Q2      2 3 4 Q2N2222
R1      6 4 5k
R2      9 1 1k
R3      2 1 1k
R4      8 7 1k
R5      5 3 1k
V1      1 0 DC 12V
V2      6 0 DC -6V
V4      5 0  AC 0.5mV
V3      8 0  AC 0.5mV
.DC LIN V3 -1 1 .1
.END
```

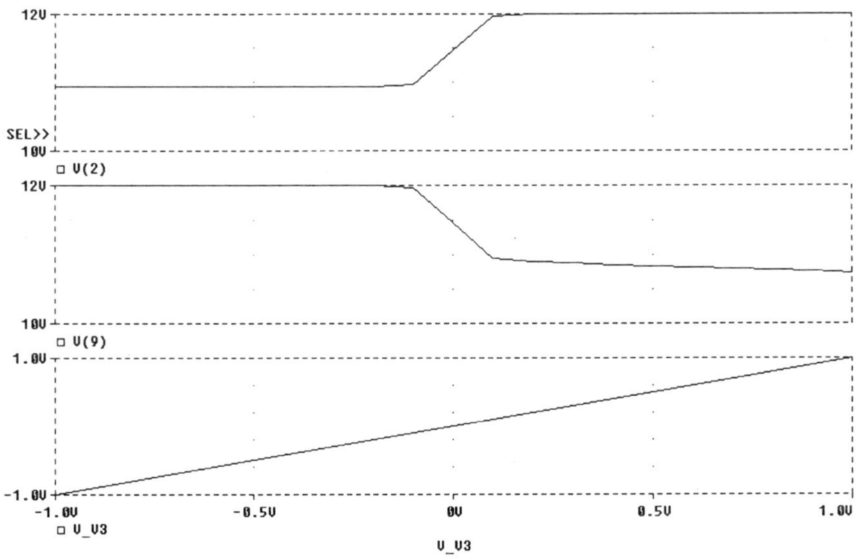

Figure 12.9 DC characteristic of the differential amplifier.

12.11 Exercises

12.1 For the following PSpice blocks determine the schematic and the type of analysis conducted.

(i)

PSpice block 12.11

```
C1      $N_0001 $N_0002 10n
C2      $N_0002 $N_0003 100n
R1      0 $N_0002 100k
R2      0 $N_0003 10k
R3      $N_0004 $N_0001 5k
V1      $N_0004 0  AC 4V
.ac LIN 100 10Hz 1megHz
.OP
```

(ii) (Note that the D1N750 is a zener diode)

PSpice block 12.12

```
Q1      $N_0002 $N_0001 Vout Q2N2222
Q2      $N_0002 $N_0003 $N_0001 Q2N2222
R1      $N_0002 $N_0003 10k
R2      $N_0004 Vout 10k
R3      $N_0004 $N_0002 100k
R4      Vout $N_0001 100k
R5      0 $N_0004 10k
D1      0 $N_0004 D1N750 2.5V
V1      $N_0002 0 DC 15
```

12.2 Refer to Section F.11 for more exercises.

Fault Modelling

13

13.1 Introduction

Figure 13.1 shows the typical phases of a development project. Testability should be build into a design well before the final testing phases. This makes the intermediate and final tests much easier (and also makes prototypes easier to test). Final system testing involves implementing a set of test specifications. These test specifications can either be generated automatically (using ECAD tools) or manually.

In general, testing involves two main areas:

- the verification of the design (does it meet the specification):
- fault finding.

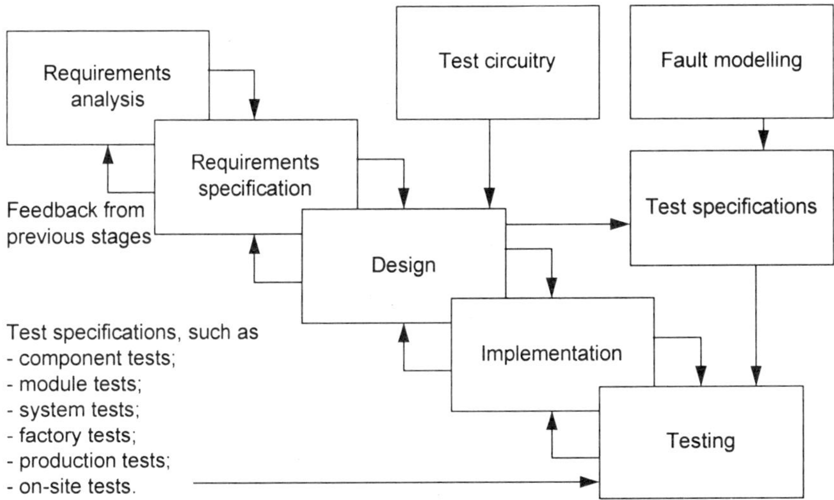

Figure 13.1 Project development.

The importance of test cannot be underestimated. A product which is faulty in the field can prove to be extremely expensive and, in many respects, embarrassing to the designers. Large electronic systems can be tested using automatic test equipment (ATE), these simulate inputs and outputs, and can also probe parts of the system. Unfortunately, in VLSI design it is difficult to probe physical parts of the circuit so it is important to add extra circuitry to a design in order to provide test information. With large-scale integration and increasing complexity it is nearly impossible to simulate all the conditions that can occur. Thus it is important to analyze the system to model the types of faults that can occur on the system and to generate a set of test vectors which will exercise these faults. This chapter discusses methods on faults modelled within a digital system and the next chapter discusses circuits which can be added to electronic circuits which aid design test.

Pattern generation and fault simulation methods are often based on simple fault models such as 'stuck-at' faults, short circuits between nodes, or the critical path algorithm. The three main parameters in test are:

- controllability;
- observability;
- predictability.

13.2 Fault coverage and fault models

Test vectors are a set of applied binary signals which have corresponding output vectors. The amount of coverage gives a measure of their testing quality, known as the fault coverage. It is often expressed as a percentage of all possible faults that the test vectors can detect. Thus:

$$\text{Fault coverage} = \frac{\text{Number of faults detected}}{\text{Total number of faults}} \times 100\%$$

Many test systems use the single-stuck-at (ssa) model to define a fault. In this model, a fault-free circuit is one in which all logic gates work properly and all interconnections (and their associated nodes) assume either logic 1 or logic 0 voltages. A faulty circuit is one which has a single gate or interconnection operating incorrectly. It is further assumed that all faults manifest themselves as if a node in the circuit was permanently held at either a 1 or a 0 level. Although the model ignores multiple faults, it is likely that in modelling single faults it also detects multiple faults.

Figure 13.2 shows the faults that can occur on a TTL gate and how they appear as stuck-at conditions. An input which is short-circuit to VCC will be stuck at a 1 and an input short-circuited to GND will be stuck at a 0. In a TTL gate an open-circuited input causes no current to flow out of the gate thus the input is taken as a high input and will thus be stuck at 1. The output can also be short-circuited to a high or a low.

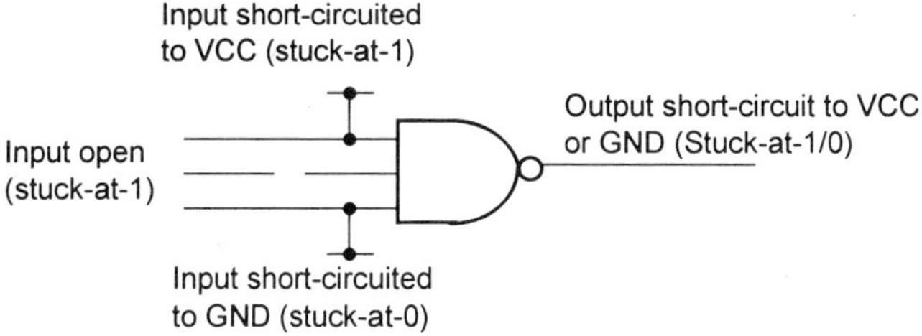

Figure 13.2 TTL stuck-at faults.

13.3 Path sensitization

One method of testing a circuit is to set up a path for a signal to propagate to the output, this is known as path sensitization. For example, if the circuit under test is a 2-input NOR gate with inputs A and B then to test the B input the A input would be set to a 0. Thus if the B input is a 0 then the output will be a 1, and if the B input is a 1 then the output will be a 0. Table 13.1 outlines this. For a 2-input NAND gate, then to test one of the inputs the other input must be set to a 1, as outlined in Table 13.2.

For an AND gate on other inputs on the gate must be a 1 for one of the inputs to propagate to the output. Whereas, an OR gate requires 0's on all of the inputs for a signal to propagate to the output. It can also be shown that the output on an exclusive-OR gate (XOR) always changes when any one of the inputs change (thus the applied inputs are don't care states). For a 2-input XOR gate, if one of the inputs is a 0 then the signal will be inverted. If one of the inputs is a 1 then the signal will be inverted on the output.

Table 13.1 NOR gate testing.

A	B	Z
0	0 ⟶	1
0	1 ⟶	0
1	0	0
1	1	0

Table 13.2 NAND gate testing.

A	B	Z
0	0	1
0	1	1
1	0 ⟶	1
1	1 ⟶	0

Figure 13.3 shows an example of a circuit to be tested. For a fault on A (that

is, a stuck at fault on N_1), then the inputs B and C must be set at a state that allows a change in A to cause a change in the output.

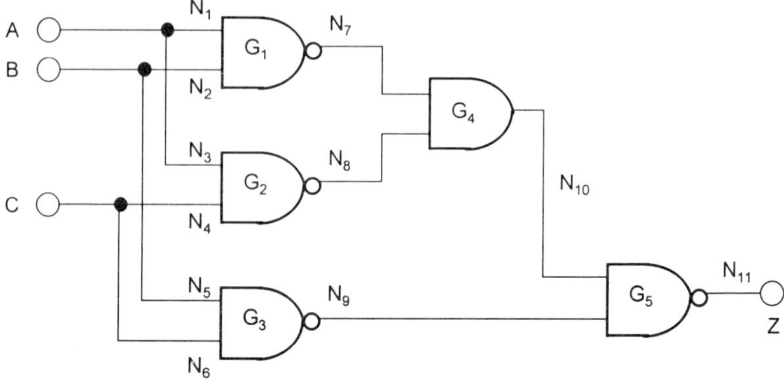

Figure 13.3 Boolean circuit to test ssa fault.

To sensitize the fault on the N_1 node the path to the output must be sensitized, that is, the path from N_1 to N_7 to N_{10} to N_{11}. This path can be traced either from the output to the tested node or from the tested node to the output. To trace forward, the N_2 node must be set to a 1 because of the NAND gate G_1. This sensitizes N_7 and the B input must thus be a 1.

To propagate the N_7 node through to N_{10} then the N_8 node must be a 1. To make this a 1 then one of the inputs into the AND gate G_2 must be a 0. Since the B input is a 1 then the N_4 node must be a 0. This makes C a 0.

To propagate the N_{10} node to N_{11} then the N_9 node must be 1. The inputs to G_3 are a 1 (B input) and a 0 (C input). Thus the output N_9 will be a 1, thus the N_{10} signal will propagate to N_{11}.

The test vector for a stuck-at zero fault on N_1 will be 110 and the test vector for a stuck-at one fault on N_1 is 010. When the A input is a 1 then the output of G_1 will be a 0 (because B is also a 1). The output of G_2 will be a 1 since A is a 1 and C is a 0. The output of G_4 will thus be a 0. The output of G_3 will be a 1 since B is a 1 and C is a 0. The output of G_5 will thus be a 1. Thus for a fault-free circuit an input of 110 should given an output of 1.

To provide that the output differs for the other test vector the output is tested for a 010 test vector. This should give an output of 0. To prove this the circuit is traced from input to output. The output of G_1 will be a 1 and the output of G_2 will also be a 1. This makes the output of G_4 a 1. The output of G_3 will be 1, thus the output of G_5 will be a 0. Figure 13.4 and Figure 13.5 show the expected logic level for each test vector and Table 13.3 summarizes the test vectors.

The sensitization of the path for the N_1 node also tests for other nodes in the path (because each of them must change when the test signal changes). Thus N_7, N_{10} and N_{11} are tested for ssa faults.

Table 13.3 Test vectors for a ssa fault on N1.

ABC	*Z*	*Type of fault tested*
110	1	Stuck-at-zero fault on N1
010	0	Stuck-at-one fault on N1

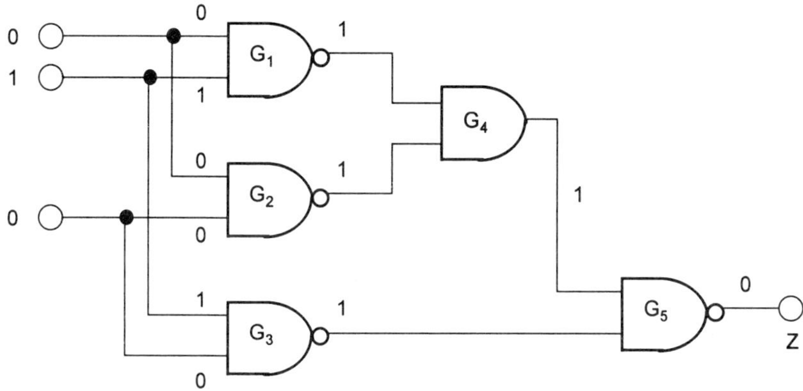

Figure 13.4 Test vector levels for 010 input.

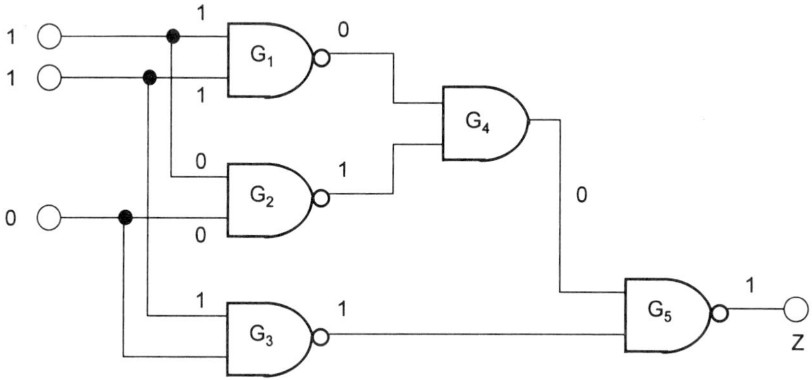

Figure 13.5 Test vector levels for 110 input.

13.3.1 Boolean differences

Path sensitization is relatively simple for small circuits but can be difficult if the circuit has many gates. To reduce this need an algebraic technique called Boolean differences can be used to develop test patterns for combinational circuits. This technique is relatively simple to implcment in software.

The Boolean difference technique defines the Boolean equation of a circuit with the sensitize input set as a 1 and also set at a 0. The other inputs are then set so that these two equations differ, that is, one is a logic 1 and the other is a 0. If the two equations are exclustive-OR'ed then the result will be a logic 1.

Described mathematically, the Boolean difference of a function Z with respect to variable input X is:

$$\frac{dZ}{dX} = Z_x(x=0) \oplus Z_x(x=1)$$

where $Z_x(0)$ is the value of Z when X=0 and $Z_x(1)$ is the value of Z when X=1.

Thus to make dZ/dX=1, then the $Z_x(0)$ and $Z_x(1)$ must differ. To develop a set of test vectors for a function with a single input stuck 1 (stuck at 0) a fault a path must be sensitized that will exercise the fault when the variable is set to 0 (1). Mathematically

$$\overline{X}\frac{dZ}{dX} = 1 \text{ will test for a stuck at 1 fault}$$

$$X\frac{dZ}{dX} = 1 \text{ will test for a stuck at 0 fault}$$

A simple example is a NOR gate,

$$Z = \overline{A+B}$$

$$Z_A(A=0) = \overline{B}$$

$$Z_A(A=1) = 0$$

Thus

$$\frac{dZ}{dA} = Z_A(A=0) \oplus Z_A(A=1)$$

$$= \overline{B} + 0$$

$$= \overline{B}$$

Thus to make dZ/dA equal to 1, B must be a 0. This confirms that a 2-input NOR gate requires one of the inputs to be a 0 in order to test the other input. Another example is to find the test vectors of a fault on the C input for the equation $Z = \overline{A}.B.C + A.\overline{B}.\overline{C} + ABC$.

$$Z_c(0) = A.\overline{B}$$

$$Z_c(1) = \overline{A}.B + AB$$

$$Z_c(0) = A.\bar{B}$$

$$Z_c(1) = \bar{A}.B + AB$$

$$\begin{aligned}
\frac{dZ}{dC} &= Z_C(0) \oplus Z_c(1) \\
&= A.\bar{B} \oplus (\bar{A}.B + AB) \\
&= A.\bar{B}.\overline{(\bar{A}.B + AB)} + \overline{A.\bar{B}}.(\bar{A}.\bar{B} + AB) \\
&= A.\bar{B}.\overline{(A + \bar{B} + AB)} + (\bar{A} + B).(\bar{A}.\bar{B} + AB) \\
&= A.\bar{B}.(\bar{A} + B + \overline{AB}) + \bar{A}.\bar{A}.\bar{B} + B.\bar{A}.\bar{B} + \bar{A}AB + BAB \\
&= A.\bar{B}.\bar{A} + A.\bar{B}.B + A.\bar{B}.(\bar{A} + \bar{B}) + 0 + 0 + A.B \\
&= 0 + 0 + A.\bar{B}.\bar{A} + A.\bar{B}.\bar{B} + A.B \\
&= A.\bar{B} + A.B
\end{aligned}$$

Thus if A=1 and B=0 or if A=1 and B=1 then the output depends on C. To test for stuck at 0 faults:

$$C\frac{dZ}{dC} = A.\bar{B}.C + A.B.C$$

Thus the test vectors are 101 and 111, and the output will be a 1. To test for stuck at 1 fault:

$$\bar{C}\frac{dZ}{dC} = A.\bar{B}.\bar{C} + A.B.\bar{C}$$

Thus the test vectors are 100 and 110, and the output will be a 0.

13.4 Simulating stuck-at faults

Many ECAD simulators allow stuck-at faults to be modelled. A useful portable and quick method is to use a high-level language to model the Boolean equations. Program 13.1 simulates the equation:

$$Z = \overline{A.\overline{B + \bar{C}}}$$

The test run 13.1 shows the output from the program. This shows the output without any faults (Z), with a fault on A stuck at 0 (ZA0), with a fault on A stuck at 1 (ZA1), and so on. It can be seen that in this case the output produced with no faults is a unique solution. It can also be seen that stuck at zero fault on A causes the output to be a 1, no matter the state of the other inputs (as happens when C is also stuck at 0).

Program 13.1

```
#include <stdio.h>

#define  TRUE  1
#define  FALSE 0

int    NAND(int x, int y);
int    NOR(int x, int y);

int    main(void)
{
int    A,B,C,Z;
int    Z_SSA_A0,Z_SSA_A1,Z_SSA_B0,Z_SSA_B1,Z_SSA_C0,Z_SSA_C1;
int    a,b,c;

   printf("   A    B    C    Z   ZA0 ZA1 ZB0 ZB1 ZC0 ZC1\n");

   for (A=0;A<=TRUE;A++)
      for (B=0;B<=TRUE;B++)
         for (C=0;C<=TRUE;C++)
         {
            a=A;b=B;c=C;
            Z=NAND(a,NOR(b,c));

            a=TRUE;b=B;c=C;
            Z_SSA_A1=NAND(a,NOR(b,c));

            a=FALSE;b=B;c=C;
            Z_SSA_A0=NAND(a,NOR(b,c));

            a=A;b=TRUE;c=C;
            Z_SSA_B0=NAND(a,NOR(b,c));

            a=A;b=FALSE;c=C;
            Z_SSA_B1=NAND(a,NOR(b,c));

            a=A;b=B;c=TRUE;
            Z_SSA_C1=NAND(a,NOR(b,c));

            a=A;b=B;c=FALSE;
            Z_SSA_C0=NAND(a,NOR(b,c));

            printf("%4d %4d %4d %4d %4d %4d %4d %4d %4d\n",A,B,C,
            Z,Z_SSA_A0,Z_SSA_A1,Z_SSA_B0,Z_SSA_B1,
            Z_SSA_C0,Z_SSA_C1);
         }
   return(0);
}

int    NAND(int x,int y)
{
   return(!(x && y));
}

int    NOR(int x, int y)
{
   return(!(x || y));
}
```

Test run 13.1

A	B	C	Z	ZA0	ZA1	ZB0	ZB1	ZC0	ZC1
0	0	0	1	1	1	1	1	1	1
0	0	1	0	1	0	1	1	1	1
0	1	0	1	1	1	1	1	1	1
0	1	1	1	1	1	1	1	1	1
1	0	0	1	1	1	1	1	1	0
1	0	1	0	1	0	1	0	1	0
1	1	0	1	1	1	1	1	1	1
1	1	1	1	1	1	1	0	1	1

13.5 Testing stuck-on/off faults in transistors

Electrical faults are normally grouped into three categories:

- open circuits - where a connection has been open-circuited;
- short circuits - where a connection has been short-circuited either to a power rail or a signal line;
- stuck-on/off faults - where a transistor is stuck-on is stuck-off.

A stuck-on faults causes a FET to remain conducting regardless of the gate input and a stuck-off fault remains off regardless of the input. Figure 13.6 shows a CMOS gate with a stuck-off fault. When the input to the NAND gate is AB=00 then the output goes high because of the pull-ups connected to V_{DD}. An input of AB=11 causes the output to go into a high impedance because both the pull-ups and the stuck-off device are OFF.

With the NMOS equivalent circuit there will be an active pull-up transistor. The same stuck-off will always give a LOW as there is no path to ground no matter the states of A and B. Table 13.4 shows the resultant truth table with the stuck-off transistor fault.

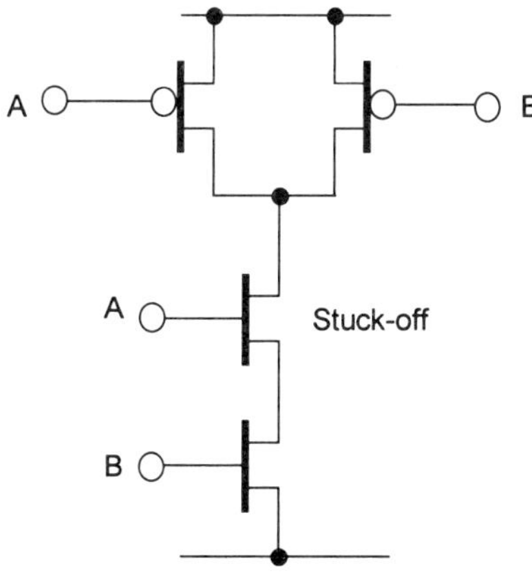

Figure 13.6 Stuck-at fault.

Table 13.4 Truth table for a NAND gate with a stuck-off NMOS transistor on A input.

A	*B*	*Correct output*	*Faulty output*
0	0	1	1
0	1	1	1
1	0	1	1
1	1	0	High impedance

Table 13.5 shows the truth table for a stuck-on NMOS transistor connected to the A input. The problem in this circuit occurs when the A input is a 0 and the B input is a 1. This is because there are two pull-down transistors on and the pull-up PMOS transistor connecting the A input is also on. Thus there is a path to the upper and lower supply rail. The actual voltage level will depend on the actual on-resistance of the transistors. If each of them has the same resistance then the output voltage will be two-thirds of the supply voltage (this level is likely to be taken as a logic high).

Stuck-at faults can also occur on the pull-up transistors. These are stated in Table 13.6 and Table 13.7.

Table 13.5 Truth table for a NAND gate with a stuck-on NMOS transistor on A input.

A	*B*	*Correct output*	*Faulty output*
0	0	1	1
0	1	1	1 (although a lower voltage output than normal high level)
1	0	1	1
1	1	0	0

Table 13.6 Truth table for a NAND gate with a stuck-off PMOS transistor on A input.

A	*B*	*Correct output*	*Faulty output*
0	0	1	1
0	1	1	High impedance
1	0	1	1
1	1	0	High impedance

Table 13.7 Truth table for a NAND gate with a stuck-on PMOS transistor on A input.

A	*B*	*Correct output*	*Faulty output*
0	0	1	1
0	1	1	1
1	0	1	1
1	1	0	1 (although a lower voltage output than normal high level)

13.6 Exercises

13.1 Using Boolean differences, develop sets of test vectors to detect all stuck-at faults for the following elementary logic gates: two- and three-input AND and OR gates, an inverter, a two-input exclusive-OR gate, and two- and three-input NAND and NOR gates. Determine the output for these tests.

13.2 Using the software developed in Program 13.1 determine the truth table with stuck-at faults for the following equations:

(a) $Z = \overline{(A.\overline{B})+(B.C)}$

(b) $Z = \overline{A.B.\overline{C}+\overline{B.D}}$

13.3 Determine a set of test vectors for a fault on input A by:

(a) sensitizing the path by reasoning;
(b) using Boolean differences.

13.4 Using Boolean differences, determine the test vectors to test for a fault on input B for the equation $Z = \overline{AB+CD}$.

13.5 The traffic light controller developed in Chapter 7 used the following equations:

$$RED\_EW = G$$

$$RED\_NS = \overline{G}$$

$$YELLOW\_EW = \overline{G}.F.E$$

$$YELLOW\_NS = G.F.E$$

$$GREEN\_EW = \overline{G}.\overline{F}+\overline{G}.\overline{E}$$

$$GREEN\_NS = G.\overline{F}+G.\overline{E}$$

$$WALK\_EW = \overline{F}.\overline{G}$$

$$DONT\_EW = G + A.F.\overline{E}+\overline{G}.F.E$$

$$WALK\_NS = G.\overline{F}$$

$$DONT\_NS = \overline{G}+A.F.\overline{E}+G.F.E$$

Determine test vectors for each of the outputs.

13.6 Design a 1-bit full adder using a minimum of AND gates, OR gates and inverters (AOI logic). Using the sensitive path techniques, develop a set of test vectors that identifies all possible stuck-at faults for this circuit.

13.7 Design a three-input majority circuit (that is, its output is a 1 if more than two of the inputs are a 1) using the minimum AOI logic. Using sensitive path techniques, develop a set of test vectors to test for all possible single stuck-at faults for this circuit.

13.8 Show that if the circuit in Figure 13.3 has the input to the G_3 is inverted that there is no test for a stuck-at 0 fault on N_1.

13.9 Design a CMOS circuit to implement the following Boolean equation:

$$Z = \overline{[(A + B) \cdot (C + (A \cdot D))]}$$

If the PMOS transistor connected to the D input is stuck ON explain how the output will differ from the expected results.

13.10 A CMOS logic device has a single stuck-at fault on a transistor. When tested it produces the results shown in Table 13.8. Determine the circuit and the transistor which has the fault. Note the supply voltage is +5V.

Table 13.8 Truth table.

A	*B*	*C*	*Output voltage*
0 V	0 V	0 V	2 V
0 V	0 V	5 V	0 V
0 V	5 V	0 V	1.7 V
0 V	5 V	5 V	0 V
5 V	0 V	0 V	1.7 V
5 V	0 V	5 V	0 V
5 V	5 V	0 V	0 V
5 V	5 V	5 V	0 V

13.11 Develop a full set of truth tables for stuck-on/off faults on a 2-input CMOS NOR gate. Assume that a stuck-on transistor has the same resistance as a transistor which is on. Also assume that the threshold for a low level is any voltage less than one-fifth the supply voltage. A voltage higher than this is taken as a high input.

13.12 A circuit which implements the function:

$$Z = \overline{(A.B) + (\overline{A}.\overline{B}.C) + (\overline{A}.\overline{B}.\overline{C}.\overline{D})}$$

has a single stuck-at fault on one of the inputs. The expected and tested results are given in Table 13.9. Locate the fault and determine at which logic state it is stuck-at.

Table 13.9 Truth table.

A	*B*	*C*	*D*	*Z(Expected)*	*Z(Tested)*
0	0	0	0	0	1
0	0	0	1	1	1
0	0	1	0	0	0
0	0	1	1	0	0
0	1	0	0	1	1
0	1	0	1	1	1
0	1	1	0	1	1
0	1	1	1	1	1
1	0	0	0	1	1
1	0	0	1	1	1
1	0	1	0	1	1
1	0	1	1	1	1
1	1	0	0	0	0
1	1	0	1	0	0
1	1	1	0	0	0
1	1	1	1	0	0

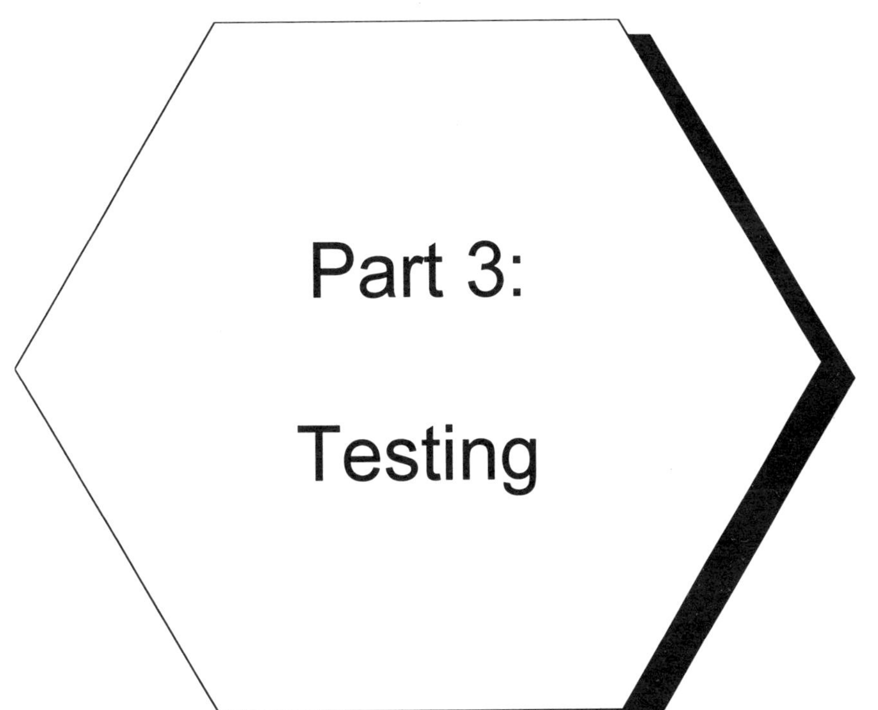
Part 3:
Testing

Designing for Test

14

14.1 Introduction

As electronic systems become more complex and more compact the need for systems which either help the test process or can even test themselves is growing. Many companies now desire zero defects in their products so it is important that defects are detected before they are reach the market. All these factors result in a need for testability to be built into a design at an early stage.

14.2 Combinational and sequential circuits

Combinational logic circuits are relatively easy to test. Input signals are applied and the outputs are compared with simulated results. Sequential circuits are more difficult to test as their output depends on both the present inputs and the internal state of the circuit. A microprocessor device is extremely difficult to test because although it may not have a great deal of inputs and outputs the number of possible input sequences is extremely large. One method used to test sequential networks is to split them up into blocks of combinational logic, which can be independently tested. Each of the combination blocks are separated by flip-flops and registers to be able to read the output state.

14.3 Scan-in scan-out design

It is easier to test a combination circuit as the current output does not depend on any previous inputs. Whenever possible a design should be partitioned into elements of combinational logic. Boundaries are then partitioned by flip-flops and latches, as illustrated in Figure 14.1. The multiplexer switches between the test inputs or the input from a previous stage. When the test signal becomes active then the test inputs are selected. The previous stage outputs from a partition can be monitored to test the input test vector against the output.

The main disadvantage of loading the test data in a parallel form is that it requires many more input/output pins and internal connections. An improved method is to load the test pattern in a serial form. Figure 14.2 shows how the test pattern can be loaded into the inputs to a partitioned circuit. When the test line is low then the 0 input of the multiplexers are selected and with the clock (CLK) the data will be loaded serially into the multiplexers. This operation converts it into a parallel form.

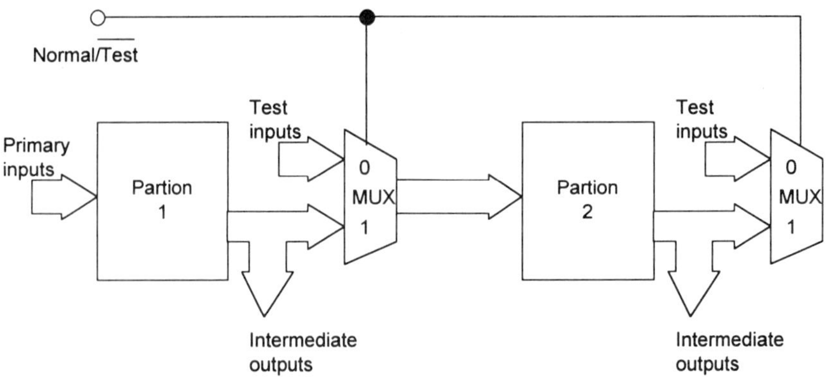

Figure 14.1 Partitioning of design.

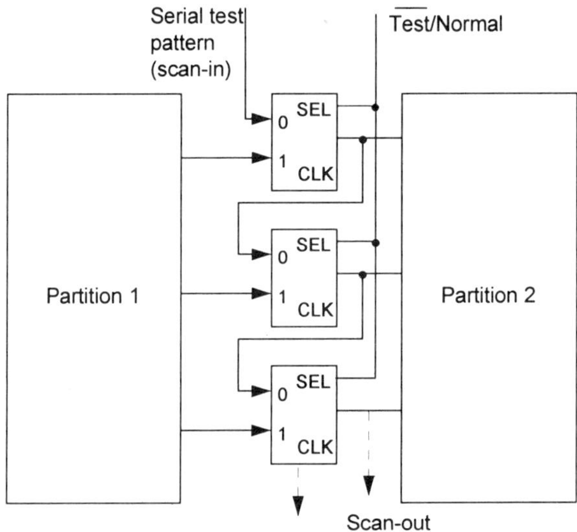

Figure 14.2 Serial input for the test pattern.

Once the test pattern has been scanned in, the registers return to their normal mode and a single clock inputs the pattern into the partitioned circuit. The test pattern then propagates through the partitioned circuit and once the circuit has settled the output of the circuit is scanned out by returning the registers into shift mode. This is then repeated for each partitioned element within the design. The input registers of one element serves as output registers for another.

The SISO method normally reduces the amount of time taken to test the device as it reduces the amount of test vectors used. It helps to test extremely complicated devices such as microprocessors. It also requires very few extra pins, requiring the addition of a scan-in line, a scan-out line, a test/normal control line and a few other clock and control inputs.

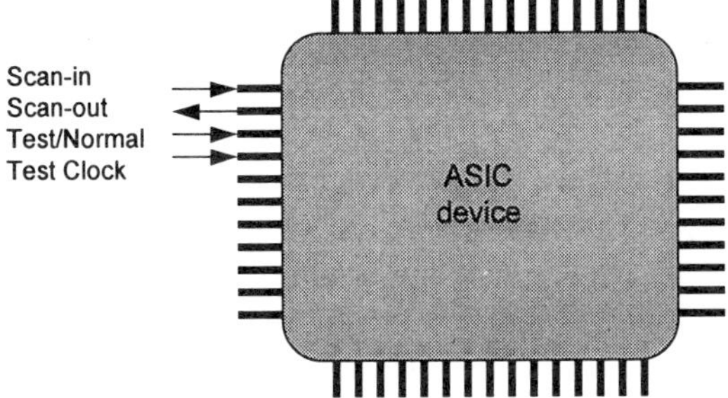

Figure 14.3 Extra pin connections required for SISO.

14.3.1 Level sensitive scan design (LSSD)

IBM developed one of the best know standards for SISO which is known as level-sensitive scan design (LSSD). When applied to on-chip self-testing it is often known as LOCST (LSSD on-chip self-test).

The main concept of LSSD requires that the operation of a circuit is independent of the dynamic characteristics of the logic elements (that is, rise and fall times, and propagation delays). It also insures that the next state of the circuit is independent of the order in which inputs change when a state change involves several inputs signals. The main features of LSSD are:

1. Latches instead of edge-triggered flip-flops are used as storage elements. A latch responds to input changes for the entire time that the control input is active, whereas edge-triggered flip-flops only respond to a rising or a falling control edge. Figure 14.4 shows a gate D-type latch made from a NAND-type SR latch which has inputs from the level sensitive CLK and an inverted and non-inverted D input. The truth table of a NAND-type SR latch is:

S	R	Q	$\overline{Q}$
0	0	Can't happen (Output oscillates)	
0	1	0	1
1	0	1	0
1	1	Q	$\overline{Q}$

It can be seen that the CLK input connects to the input NAND gates G_1 and G_2. If this CLK is a 0 then the output of G_1 and G_2 is always a 1, no matter of the other inputs. The input to G_3 will thus be a 1 and $\overline{Q}$, the output will thus be Q (the output will not change state). When the CLK is a 1 then the D input is input to the S input of the SR flip-flop and $\overline{D}$ is input on the R input of the SR flip-flop. If the D input is a 1 then the input to the SR flip flop will be SR=10 and the output will be a 1. If the D input is a 0 then the input to the SR flip-flop will be SR=01 and the output will be a 0. The outputs will change with the inputs whenever the CLK input is 1.

The level-sensitive property of latches improves timing constraints on the inputs. Thus inputs have the entire clock pulse to settle. This makes the latch operation comparatively insensitive to degradations of the edges of the clock waveforms.

2. Long, continuous shift registers are made from cascaded interconnected latches to form a long, continuous shift register to scan-in test patterns and scan-out test results. This means that all LSSD latches are implemented with a *shift-register latch* (SRL), as illustrated in Figure 14.5.

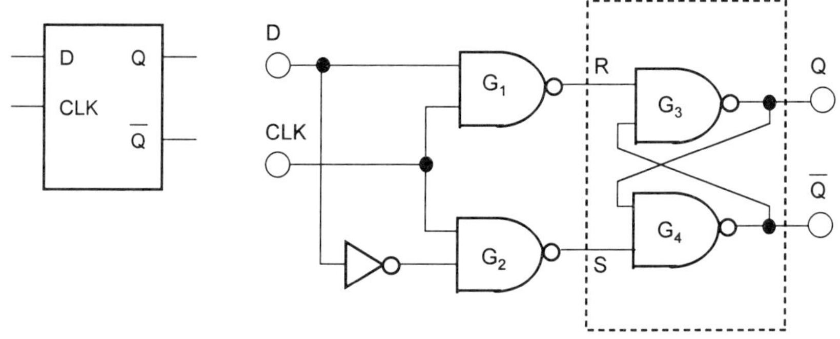

Figure 14.4 Gated D-type latch.

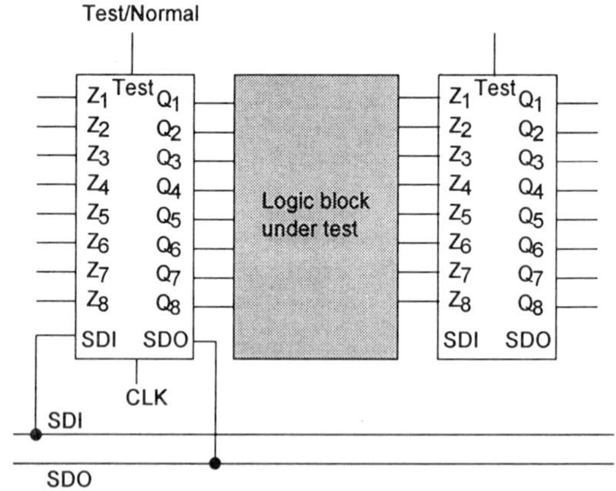

Figure 14.5 Partitioning with either a shift register latch or parallel load.

14.4 Automatic test circuits

Scanning techniques require some control to identify each of the partitioned blocks and sophisticated control mechanisms to scan the test pattern in and the results out. An improved method is to use a special circuit that allows the circuit to test itself. The output code will allow the device to either pass or fail, or can even be used to give an indication of the fault.

Self-test methods use different techniques to the traditional strobed input patterns and expected outputs. Instead, a set of input patterns are used and the outputs are compressed to reduce the storage requirements. These compressed results can be compared with the expected results.

14.4.1 Store and generate method

The simplest input test pattern is one that applies all possible input patterns to the circuit. These are stored in a memory and a binary counter simply sequences through the input test patterns. The outputs from the tested circuit are compared with expected results. This type of testing has the advantage that it is relatively simple to generate the inputs and outputs and to load them into an area of memory. A compactor can be used to reduce the number of bits on the output, as illustrated in Figure 14.6. Unfortunately this technique it is only feasible with blocks with few inputs, but it becomes time consuming for large blocks. An alternative is to apply a random sequence of input patterns, which can generate a pseudo-random binary sequence generator. This is illustrated in Figure 14.7. However, one difficulty with this approach is in determining the number of pseudo-random sequences necessary to insure good fault coverage. Problems can also arise when the partitioned blocks contain sequential elements; a random sequence of inputs may place these circuits into some 'hang' states which may take some time to exit from.

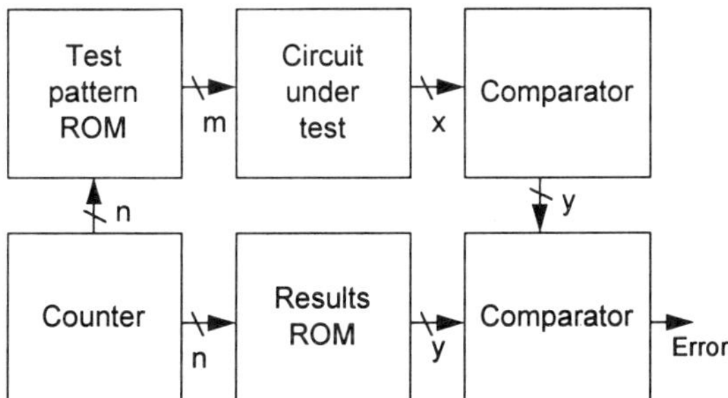

Figure 14.6 Automatic in-circuit testing.

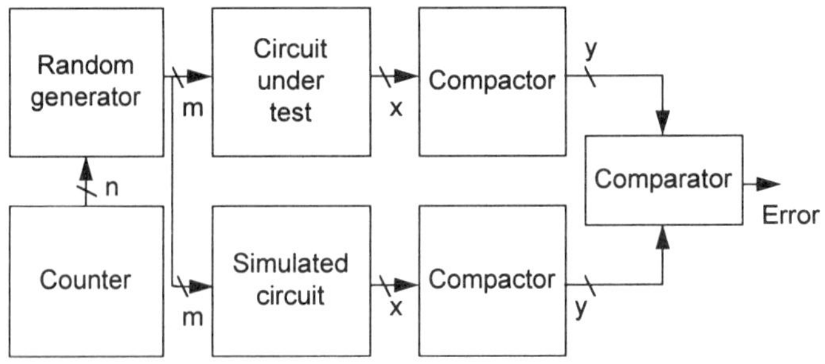

Figure 14.7 Automatic in-circuit testing with a random generator.

14.4.2 Signature analysis

Signature analysis is the most popular method of generating random test sequences and compacting output data. The main element of it is a linear feedback shift register (LFSR). Figure 14.8 shows an LFSR with the output of some stages fed back to the input through exclusive-OR gates. The output of such a shift register is a pseudo-random binary sequence. By choosing different feedback points, the length and composition of the sequence can be changed. The maximum-length sequence that can be generated by an N-stage LFSR is 2^N-1 (the all-zero state is never entered).

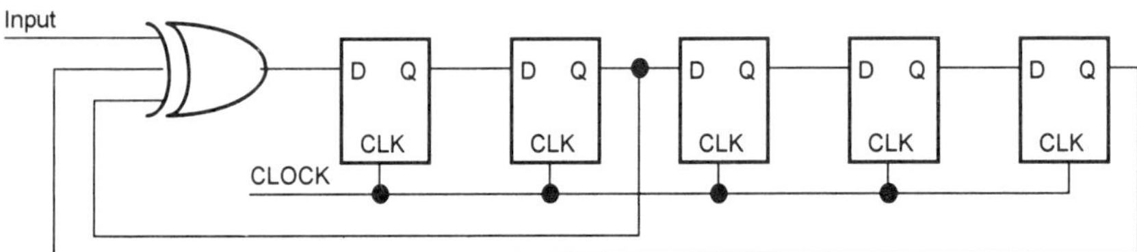

Figure 14.8 5-stage LFSR which generates a maximal-length sequence.

The output data of a tested circuit can be compacted by inputting the output of the circuit into exclusive-OR gate of the LFSR. If the input to LFSR is always a logic 0 while the register is clocked, the LFSR cycles through its states normally. If any of the clock inputs to the LFSR is a 1, then the LSFR sequence is disturbed and the actual final state is different from the expected final state. The main advantage of this circuit is that every possible digital pattern results in a different final state. It is this property that permits data compaction. If the expected sequence of data from the block being tested is known, the expected final state of the LFSR can be computed in advance. Any error in the logic blocks output disturbs the sequence and results in an incorrect final state. Note that there is no need to store and compare every expected output from the block. Only the final state needs to be compared.

Figure 14.9 shows a 5-stage LFSR with feedback from the output of the second and fifth stage. Table 14.1 shows the first eight sequences for this circuit with an input of 100101. Initially all the registers will have a zero output. Initially the C_0 will give a 1 (because the first input bit is a 1). On the next clock pulse the input state is a 0 thus C_0 will be a 0. C_1 will be a 1 because the C_0 value previous to the clock is clocked through to C_1. This continues until the required number of clock pulses.

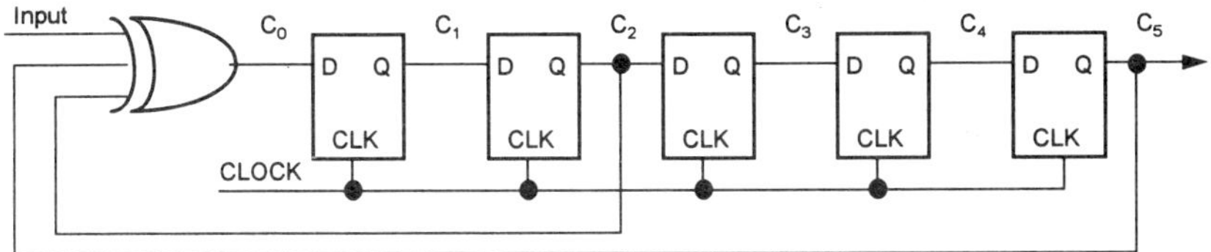

Figure 14.9 5-stage LFSR example.

Table 14.1 Sequence for circuit in Figure 14.9 with an input of 100101.

Sequence	*Input*	C_0	C_1	C_2	C_3	C_4	C_5
1	1	1	0	0	0	0	0
2	0	0	1	0	0	0	0
3	0	1	0	1	0	0	0
4	1	1	1	0	1	0	0
5	0	1	1	1	0	1	0
6	1	1	1	1	1	0	1
7	0	1	1	1	1	1	0
8	0	0	1	1	1	1	1
	and so on.						

14.4.3 LSFR program

Typically the signature for the circuit is determined using software. Program 14.1 uses the LFSR in Figure 14.9 with an input sequence of 100101 (in the sequence: 1, followed by 0, by 0, by 1, by 0 and finally 1. Test run 14.1 shows a sample run.

Program 14.1

```
/*************************************************************/
/* LFSR.C                                                    */
/* Title: Linear Feedback Shift Register                     */
/* Function: Program to display the stored states            */
/* and output sequence for an N-stage LFSR                   */
/* Author(s): Bill Buchanan                                  */
/* Version: 1.00                                             */
/* Created: 11-APR-94                                        */
/* Recent Modifications: 10-AUG-96: Bug fix in get_int()     */
/*************************************************************/
```

```
#include <stdio.h>

#define  MAX_STATES  20
#define  MAX_STAGES  20
#define  FIRST_TAP   2
#define  SECOND_TAP  5
#define  TRUE        1
#define  FALSE       0

void  calc_state(int seq,int stages,int in[],int st[],
            int out[]);
void  init_seq(int values,int st[],int in[],int ou[]);
int   X_OR(int a,int b, int c);
void  get_input(int in[]);
void  show_store(int seq,int stages,int st[]);
void  show_sequence(int vals,int arr[]);
void  start_up_screen(void);
void  get_parameters(int *no_stat,int *no_stag);
void  get_int(char msg[],int min,int max,int *val);

int   main(void)
{
int   input[MAX_STATES],output[MAX_STATES],store[MAX_STATES];
int   seq,no_stages,no_states;

   start_up_screen();
   init_seq(MAX_STATES,store,input,output);
   get_parameters(&no_states,&no_stages);
   get_input(input);
   for (seq=0;seq<no_states;seq++)
   {
      calc_state(seq,no_stages,input,store,output);
      show_store(seq,no_stages,store);
   }
   printf("INPUT  >>");show_sequence(no_states,input);
   printf("OUTPUT >>");show_sequence(no_states,output);
   return(0);
}

void  init_seq(int values,int st[],int in[], int ou[])
{
int   i;

   for (i=0;i<values;i++)
   {
      st[i]=0;
      in[i]=0;
      ou[i]=0;
   }
}

void  get_parameters(int *no_stat,int *no_stag)
{
   get_int("Enter number of states required >>",1,MAX_STATES,no_stat)
   get_int("Enter number of stages required >>",1,MAX_STAGES,no_stag)
}

void     get_input(int in[])
{
   /* setup fixed data input of 100101 */
   in[0]=1;in[1]=0;
   in[2]=0;in[3]=1;
```

```
   in[4]=0;in[5]=1;
}

void  calc_state(int seq,int stages,int in[],int st[],int out[])
{
int   i;

   st[0]=X_OR(in[seq],st[FIRST_TAP],st[SECOND_TAP]);

   for (i=stages;i>0;i--)
   {
      st[i]=st[i-1];
   }
   out[seq]=st[stages];

}

void  show_store(int seq,int stages,int st[])
{
int   i;
   printf("State %2d >> ",seq);

   for (i=1;i<=stages;i++)
      printf("%d",st[i]);
   printf("\n");
}

void  show_sequence(int vals,int arr[])
{
int   i;
   for (i=0;i<vals;i++)
      printf("%d",arr[i]);
   printf("\n");
}

int   X_OR(int a,int b, int c)
{
   return((a^b^c));
}

void  start_up_screen(void)
{
   puts("");
   puts("\t***********************************");
   puts("\t* LFSR.EXE    Version 1.00        *");
   printf("\t* Date last modified %s   *\n",__DATE__);
   printf("\t* Program author: Bill Buchanan  *\n");
   puts("\t***********************************");
}

void  get_int(char msg[],int min,int max,int *val)
{
char  instr[BUFSIZ];
int   rtn,okay;

   do
   {
      printf("%s",msg);
      gets(instr);
      rtn=sscanf(instr,"%d",val);
      if ((rtn!=1) || (*val<min) || (*val>max))
      {
```

```
            okay=FALSE;
            printf("Invalid input <%s>. Valid values are %d to %d\n",
                                                  instr,min,max);
        }
        else okay=TRUE;

    } while (!okay);
}
```

Test run 14.1

```
        ***********************************
        * LFSR.EXE      Version 1.00      *
        * Date last modified Aug 25 1996  *
        * Program author: Bill Buchanan   *
        ***********************************
Enter number of states required >>15
Enter number of stages required >>6
State  0 >> 100000
State  1 >> 010000
State  2 >> 101000
State  3 >> 110100
State  4 >> 111010
State  5 >> 111101
State  6 >> 111110
State  7 >> 011111
State  8 >> 001111
State  9 >> 100111
State 10 >> 110011
State 11 >> 011001
State 12 >> 101100
State 13 >> 010110
State 14 >> 001011
INPUT  >>100101000000000
OUTPUT >>000001011111001
```

14.4.4 Built-in logic block observers (BILBO)

Linear-feedback shift registers can be used for both random input generation and output data compression to create what is called built-in logic block observer (BILBO). A BILBO is a standard structure which can be used to load inputs in parallel (in the normal mode), to generate a random sequence or can be used for signature analysis. The logic diagram of an 8-bit BILBO is shown in Figure 14.10. It shows that the two control lines are A and B. These set up the four control modes:

- parallel load (normal mode);
- synchronous clear;
- pseudo-random generator;
- signature analysis.

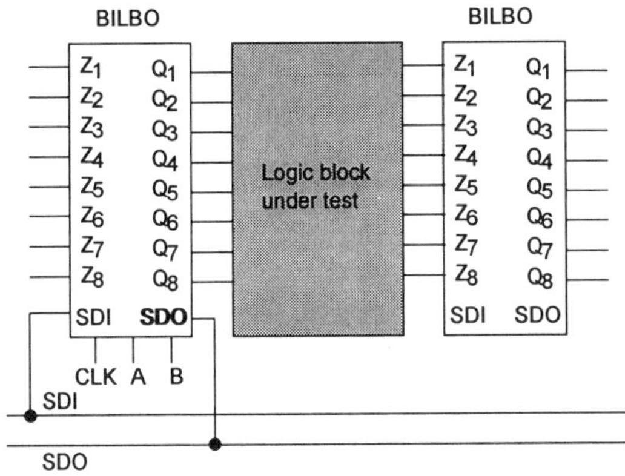

Figure 14.10 An 8-bit BILBO circuit.

When testing a circuit the circuit is placed in its test mode and the BILBOs are initialized. The circuit is then clocked through the entire pseudo-random sequence (255 clock sequences for an 8-stage LFSR) to generate a circuit signature. After the final clock cycle, the contents of the signature BILBO are compared with the expected signature. A different signature indicates a circuit fault.

The logic control block within the BILBO is illustrated in Figure 14.11. When A is a 1 then the output of the BILBO control block exclusive OR (G_2) will always be a zero. Thus the output of the exclusive-OR gate (G_2) will be a 0, no matter the state of the other input to G_2. The output from the and gate (G_1) will thus be mirrored on the output of the G_2, which is fed into the D-type flip-flop. If the B input is a 0 then the output of G1 will always be a 0, thus the input to all the D-type flip-flop will be 0s. This is the synchronous clear mode (AB=10). If the B input is a 1 then the Z inputs (Z_1 to Z_8) will appear on the inputs to each stage. This mode is the parallel load (AB=11) and is the normal mode.

When A is a 0 then the output of the G_2 will be the inverse of the output of the feedback of the outputs (Q_2, Q_4, Q_5 and Q_7) fed through the exclusive-OR gate (G_4). If B is a 1 then the input will be taken from the Z inputs. This mode is the signature generator mode (AB=01). If the B input is a 0 when A is a 0 then the output of G_2 is exclusive-OR'ed with 0. This mode is the pseudo-random generator mode. Table 14.2 lists the modes.

Table 14.2 BILBO operations.

A	*B*	*Operation*
1	1	Parallel load from inputs Z_1 to Z_8 (normal operation)
1	0	Synchronous clear
0	0	Generate pseudo-random sequence (test mode 1)
0	1	Generate parallel signature (test mode 2)

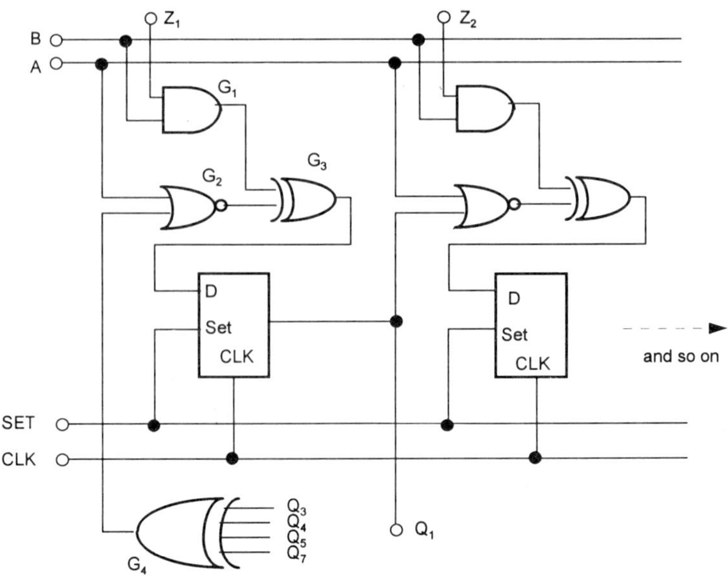

Figure 14.11 Simplified BILBO schematic.

14.5 Exercises

14.1 Discuss the Scan-In Scan-Out method of testing circuits showing how a digital circuit can be modified to incorporate it.

14.2 Discuss how a built-in logic block observer (BILBO) circuit can be used to test a digital circuit.

Memory Tests

15

15.1 Introduction

Chapters 13 and 14 discuss electronic testing in some detail. This section outlines basic test patterns for testing RAMs. There are five main types of failures involved with RAM.

1. Decoder failures – where whole potions of memory are inaccessible because of faults in the row and column decoders.
2. 'Sleeping sickness' – is the failure of a memory cell to retain its data for the specified refresh time.
3. Pattern sensitivity – is when incorrect data storage occurs only when the certain patterns are loaded into memory. It often occurs when there is interaction between physically adjacent memory cells.
4. Multiple writes – where writing to one location causes data to be written to one or more other locations.
5. Slow sense amplifier recovery – this is a tendency for the sense amplifiers to favour one logic value after reading a long sequence of that value.

15.2 Test patterns

It is not possible to test all the possible sequences that can be applied to a large memory array. There are a number of standard test patterns which can test for the most common faults.

15.2.1 Marching 1's

The algorithm for the marching 1's is as follows:

```
initialize memory with all 0's;
repeat
    read a 0 from cell and verify;
    write a 1 to the cell;
    move onto next cell;
until end of array;

repeat
```

read a 1 from cell and verify;
write a 0 to the cell;
move on to previous cell;
until start of array;

Figure 15.1 shows a sample of the marching 1's pattern.

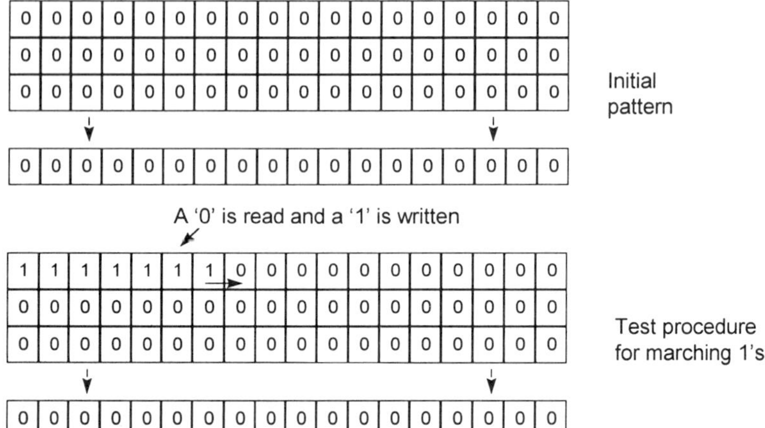

Figure 15.1 Marching 1's pattern.

For an n-bit array the total number of operations will be:

	Total operations
Initialize array	n
Read a zero and verify	2×n
Write a 1	n
Read a 1 and verify	2×n
Write a 0	n
Total	6×n

15.2.2 Marching 0's

The matching 0's pattern is identical to the marching 1's but pattern is inverted. Marching patterns detect the following:

- multiple writes;
- cells stuck at a given logic value.

A multiple write to a memory location above the address location is detected as the erroneous cell when the cell is initially read-from. On the first phase all cells should be 0 when read. If it is not then a write to previous cell has caused a 1 to be write to the current cell, or the cell is stuck at a 1.

In the second phase each cell is initially read for a 1. If it is not, then the cell is either stuck at a 0 or a write to a previous cell has caused a multiple write.

The test has the advantage of being relatively fast, the total bit being of the order of the number of bits in the array (6n). This test pattern has many disadvantages including:

- It does not exercise the address decoder. The marching patterns are sequential; most changes only alter a few bits. Consequently, internal noise is relatively low.
- They are very weak for detection of pattern sensitivity or slow sense amplifier recovery.

15.2.3 Checkerboard

The checkerboard test loads the memory with alternative 0's and 1's. This pattern is then verified. The inverse pattern is applied and again it is verified. It requires only 4×n operations (one to write then read, then the inverse). Figure 15.2 shows a sample pattern.

The checkerboard pattern sets neighbouring cells to an opposite state so it is useful in detecting shorts between adjacent cells. A possible problem with the checkerboard test pattern is that memory arrays tend not to be arranged as their logical layout. For example row 1, column 2 may not be directly above row 2, column 2. This is because memory arrays are designed to minimize cross-overs and signal delays. To overcome this problem the pattern written to the array should be a true physical checkerboard and not a logical checkerboard. This requires that the physical layout of the array is known.

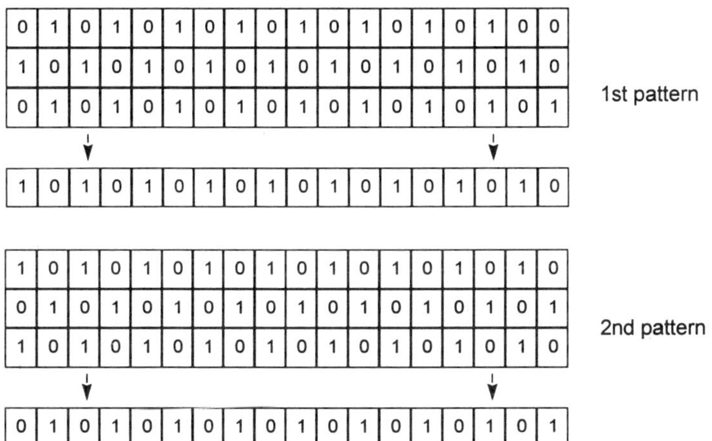

Figure 15.2 Checkerboard patterns.

15.2.4 Shifted diagonal test

The shift diagonal test is an extensive test which is used to test the sense amplifier recovery. Figure 15.3 shows a sample pattern and the algorithm is given next:

load memory with all 0's except for 1's along the diagonal;
repeat
  read each cell in the array one at a time moving down columns;
  diagonal moved 1 column to the right;
until last diagonal;

repeat with inverse pattern;

An n-bit array which is arranged in a $\sqrt{n} \times \sqrt{n}$ will have the total number of operations:

	Total operations
Read array	$n\times\sqrt{n}$
Write array	$n\times\sqrt{n}$
Total	$2\times n\times\sqrt{n}$

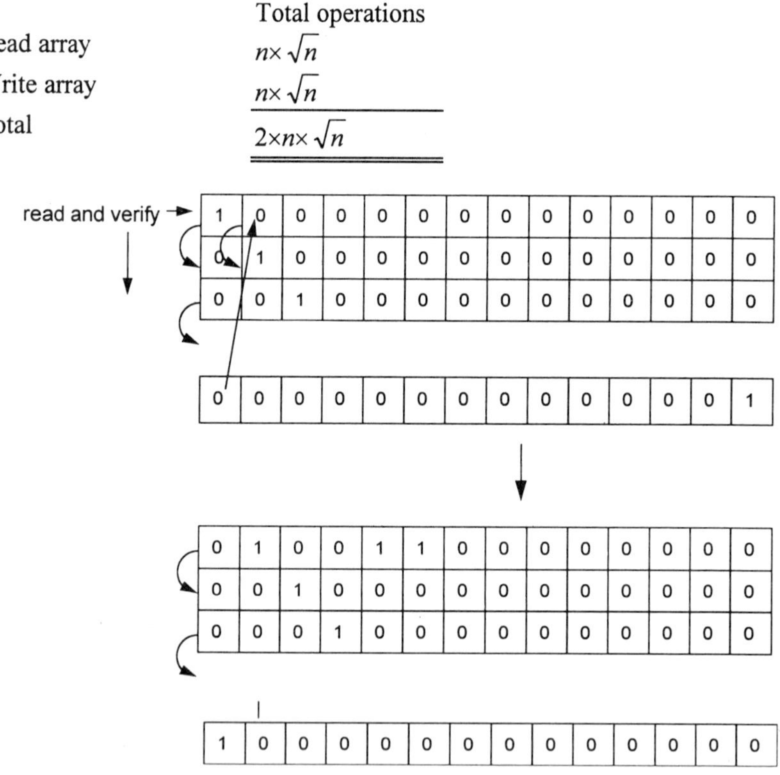

Figure 15.3 Shifted diagonal pattern.

15.2.5 Walking 1's (WALKPAT)

The walking patterns are more stringent than marching patterns. They basically involve writing to a cell, reading its contents and then rewriting its original state.

```
write 0's to all cells;
repeat
    verify that all cells in the array are 0's;
    write a 1 to the current cell;
    verify it is a 1;
    rewrite a 0;
    go to next cell;
until end of array;
repeat algorithm with the inverse;
```

An n-bit array the total number of operations (for one half of the algorithm) is approximately:

	Total operations
Verify array	$n \times n$
Write/read/write	$3\times n$
Total	$n\times n+3\times n$

The WALKPAT pattern demonstrates that each cell is addressable and that each cell can be set to a 0 and a 1. It also shows that writing to a cell does not disturb any other cell. Unfortunately it uses sequential addressing and thus does not generate much noise with the array.

As has been shown, the number of operations varies approximately with the square of the number of bits in the array. For example a 16 Mbit array (16 777 216 bits) would take at least 2.8×10^{14} operations. With a clock rate of 100 MHz, this would take 2 814 749 seconds to test. A 16 Mbit array, with a 100 MHz clock, takes 10.5 ms to read the every bit of the array. To complete the test this array must be read 16 777 216 times.

15.2.6 Galloping patterns

Galloping patterns try to test for all possible input address and output data transitions, creating a more adverse noise environment. GALPAT 1 proceeds just as a WALKPAT except that after the 1 is stored in the first cell and while the other 0's are being checked, the first cell is rechecked after each 0 is read, to ensure that the 1 remains undisturbed.

```
var     scanned cell, verify cell;
    write 0's to all cells;
```

repeat
write a 1 to the verify cell;
goto start of the array for scanned cell;
repeat
read from scanned cell;
verify it is a 0;
verify that current cell is a 1;
goto next scanned cell;
until end of array;
goto next cell;
write a 0 to the verify cell;
until end of array;
repeat algorithm with the inverse;

15.2.7 Surround disturb patterns

The walking and galloping patterns are far too slow for large memories. A good compromise (especially for DRAM memories) are the surround disturb patterns. These are designed to perform a good test for interference between neighbouring cells. The basic algorithm is:

1. write a 0 in cell m,n–1 (that is, row m, column n-1);
2. write a 0 in cell m,n;
3. write a 0 in cell m,n+1;
4. write a 1 in cell m–1,n;
5. verify that cell m,n+1 is a 0;
6. write a 1 in cell m+1,n;
7. verify that cell m,n–1 is a 0;
8. verify that cell m,n is a 0;

   repeat with complementary data.

15.3 Memory test program for a PC

Program 15.1 can be used to test 256 bytes of a computers memory. An area of 256 bytes of memory is initially allocated to the program with the `malloc()` function. A `for()` loop is then used to step through each of the 256 address locations. The `ptr++` instruction at the end of the loop increments the memory pointer by one byte because `ptr` is declared as a `char`. Within the `for()` loop eight 0's are written to the currently addressed location (`*ptr=0x00`) and then this location is read from and its contents displayed to the screen (`printf("%x",*ptr)`). Next eight 1's are written to the location and the contents are again displayed to the screen. A sample run is shown in Test run 15.1. Note that a `0x` in front of a value in C represent a hexadecimal value (thus eight 1's are represented by `0xff`) and `%x` displays a value in hexadecimal format.

Program 15.1

```
#include <stdio.h>
#include <alloc.h>

int   main(void)
{
char     *ptr;
int      i;

   puts("Program to fill memory with all 8 0's in a memory");
   puts("location then read it. Next write 8 1's into memory");
   puts("location and read it. Then move onto the next location");

   ptr=(char *) malloc(256*sizeof(char)); /* 256 bytes */

   for (i=0; i<256;i++)
   {
      *ptr=0x00;               /* write 8 0's to the current address */
      printf("(%x)",*ptr);     /* display contents of memory         */
      *ptr=0xff;               /* write 8 1's to the current address */
      printf("(%x) ",*ptr);    /* display contents of memory         */
      ptr++;                   /* goto next address                  */
   }
   return(0);
}
```

Test run 15.1

```
Program to fill memory with all 8 0's in a memory
location then read it. Next write 8 1's into memory
location and read it. Then move onto the next location
(0)(ffff) (0)(ffff) (0)(ffff) (0)(ffff) (0)(ffff) (0)(ffff)
(0)(ffff) (0)(ffff) (0)(ffff) (0)(ffff) (0)(ffff) (0)(ffff)
(0)(ffff) (0)(ffff) (0)(ffff) (0)(ffff) (0)(ffff) (0)(ffff)
(0)(ffff) (0)(ffff) (0)(ffff) (0)(ffff) (0)(ffff) (0)(ffff)
   ::::
   ::::
(0)(ffff) (0)(ffff) (0)(ffff) (0)(ffff) (0)(ffff) (0)(ffff)
(0)(ffff) (0)(ffff) (0)(ffff) (0)(ffff) (0)(ffff) (0)(ffff)
(0)(ffff) (0)(ffff) (0)(ffff) (0)(ffff) (0)(ffff) (0)(ffff)
(0)(ffff) (0)(ffff) (0)(ffff) (0)(ffff) (0)(ffff) (0)(ffff)
(0)(ffff) (0)(ffff) (0)(ffff) (0)(ffff) (0)(ffff) (0)(ffff)
(0)(ffff) (0)(ffff)
```

15.4 Exercises

15.1 Contrast the different memory test techniques.

15.2 Modify Program 15.1 so that 0's are written to 256 bytes in memory and then this is read back and displayed to the screen. The program should then repeat this for all 1's.

15.3 If possible, modify Program 15.1 so that the program tests a large area of memory (such as 64 KB, 128 KB, 256 KB, 512 KB and 640 KB). Determine the time taken for the program to test these areas of memory. Note that if a PC is being used then if more that 64 KB is to be addressed then modify the program so that the pointer is declared as a `far` pointer. A sample program for a PC which test 128 KB of data is shown in Program 15.2.

Program 15.2

```
#include <stdio.h>
#include <alloc.h>

int   main(void)
{
char     far *ptr;
int      i;

   puts("Program to fill memory with all 8 0's in a memory");
   puts("location then read it. Next write 8 1's into memory");
   puts("location and read it. Then move onto the next location");

   ptr=(char far *) malloc(128*1024*sizeof(char)); /* 256 bytes */

   for (i=0; i<256;i++)
   {
      *ptr=0x00;                /* write 8 0's to the current address */
      printf("(%x)",*ptr);      /* display contents of memory         */
      *ptr=0xff;                /* write 8 1's to the current address */
      printf("(%x)  ",*ptr);    /* display contents of memory         */
      ptr++;                    /* goto next address                  */
   }
   return(0);
}
```

15.4 Modify Program 15.1 to implement the following test patterns:

(a) Marching 1's;
(b) Marching 0's;
(c) Walking 1's.

In C the shift left operator is << and the shift right operator is >>. Thus the following routine shifts a bit position one place to the right each time:

```
val=0x01;
for (bit=0;bit<8;bit++)
{
   *ptr=val;   /* write the values 00000001, 00000010,   */
   val=val<<1; /* 00000100, and so on                    */
}
```

MOS Characteristics

A.1 MOS equations

MOS transistors are transconductance devices, where a change in the gate-source voltage causes a change in the drain-source current. In a bi-polar transistor a change in the base-emitter current causes a change in the collector current. No conduction occurs in an enhanced mode NMOS device until a threshold voltage is reached, that is, until the gate-source voltage is greater than a given threshold. Typically, this threshold is around 1 V for a 5 V supply voltage. The device has three main regions: non-conduction, linear and staturated. Equations to define the drain-source current can be derived to give:

$$I_{DS} = 0 \qquad V_{GS} < V_{TH}$$

$$I_{DS} = \frac{\beta}{2}(V_{GS} - V_{TH})^2 \qquad 0 < V_{GS} - V_{TH} < V_{DS}\,(saturated)$$

$$I_{DS} = \beta(V_{GS} - V_{TH} - \frac{V_{DS}}{2})V_{DS} \qquad V_{GS} - V_{TH} > V_{DS}\,(linear)$$

where $\beta = \frac{\mu\varepsilon}{T_{ox}} . \frac{W}{L}$

μ	-	average mobility of the charge carrier (electron for n-channel, hole for p-channel);
ε	-	the permitivity of the oxide material;
T_{OX}	-	the thickness of the oxide (layer between gate and substrate);
V_{GS}	-	gate-source voltage;
V_{DS}	-	drain-source voltage.

A typical value of the oxide layer is 700 Å (70 nm) and the electron mobility is typically $8x10^{-2}$ m^2/volt.sec. The relative dielectric constant (ε_r) of silicon oxide is approximately 4 and the permittivity of free space (ε_o) is 8.85×10^{-12} F/m.

Program A.1 determines the I-V characteristics of a FET. It uses nested loops to vary V_{GS} and V_{DS}. V_{GS} controls the outer loop and V_{DS} the inner. Initially, V_{GS} is set to V_{TH}, the inner loop starts with an initial condition of V_{DS}

=0 then is increment by 0.5 V each time round the inner loops. After the inner loops are complete the program will return to the outer loop which will then increment V_{GS} by 1 V. This continues until V_{GS} equals the supply voltage.

Program A.1

```
/* Program which will determine the drain current in a FET          */
/* with an entered threshold voltage, beta and Supply voltage      */
#include <stdio.h>
int      main(void)
{
float    Id,Vgs,Vth,VDD,Vds,beta;
   /* Id is the drain current,Vgs is the gate-source voltage      */
   /* Vth is the threshold voltage of the FET, VDD the supply     */
   /* voltage  and beta is a constant for the FET                 */
   if ((out=fopen("out.dat","w"))==NULL)
   {
      printf("Cannot open file");
      return(1);
   }
   printf("Enter Supply voltage and threshold>>");
   scanf("%f %f",&VDD,&Vth);
   printf("Enter beta (A/V^2)>>");
   scanf("%f",&beta);

   for (Vgs=Vth;Vgs<=VDD;Vgs++)
            /* gate-source voltage incremented in 1V steps */
   {
      printf("Vgs=%6.2f\n",Vgs);
      for (Vds=0;Vds<(Vgs-Vth);Vds+=0.5)
            /* drain-source voltage incremented in 0.5V steps */
      {
         Id=beta*Vds*(Vgs-Vth-0.5*Vds);
         fprintf(out,"%6.2f %6.2f\n",Vds,Id);
      }
      for (Vds=Vgs-Vth;Vds<VDD;Vds+=0.5)
      {
         Id=0.5*beta*(Vgs-Vth)*(Vgs-Vth);
         fprintf(out,"%6.2f %6.2f\n",Vds,Id);
      }
   }
   fclose(out);
   return(0);
}
```

The capacitance of a parallel plate capacitor is given by:

$$C = \frac{\varepsilon A}{d} \quad \text{Farads}$$

where A is the area of the plates and d is the distance between the plates. The gate-source capacitance can thus be found by:

$$C_G = \frac{\varepsilon}{t_{OX}}(WL) \quad \text{Farads}$$

where T_{ox} is the thickness of the oxide layer. A program to determine the gate capacitance is given in Section A.5.

A.2 NMOS invertor with depletion load

Resistors are difficult to fabricate on silicon as they require a relatively large area. An improved method is to use a depletion-mode transistor as a pull-up resistor is shown in Figure A.1. The depletion mode transistor conducts when the gate-source voltage is zero.

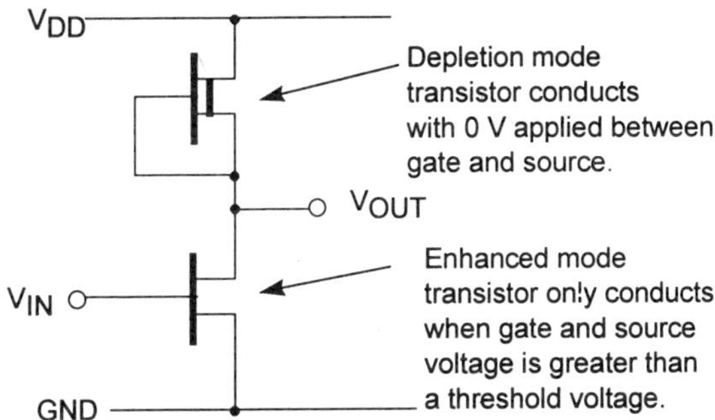

Figure A.1 NMOS inverter with pull-up load.

The equations for each transistor will be:

$$I_{PD} = 0 \qquad V_{IN} < V_{TH}$$

$$I_{PD} = \frac{\beta_{PD}}{2}(V_{IN} - V_{TH})^2 \qquad 0 < V_{IN} - V_{TH} < V_{OUT}\ (saturated)$$

$$I_{PD} = \beta_{PD}(V_{IN} - V_{TH} - \frac{V_{OUT}}{2})V_{OUT} \qquad V_{IN} - V_{TH} > V_{OUT}(linear)$$

the pull-up depletion mode transistor has a negative threshold voltage (approximately between $-0.6V_{DD}$ and $-0.8V_{DD}$) and its V_{GS} is always 0, so that $V_{GS}-V_{DEP} > 0$ and the pull-up always conducts. Its source-drain current will be given by:

$$I_{PU} = \frac{\beta_U}{2}(V_{DEP})^2 \qquad |V_{DEP}| \le V_{DD} - V_{OUT}(saturated)$$

$$I_{PU} = \beta_U(|V_{DEP}| - \frac{V_{DD} - V_{OUT}}{2})(V_{DD} - V_{OUT}) \qquad |V_{DEP}| > V_{DD} - V_{OUT}(linear)$$

In the LOW condition the pull-down transistor will be in the linear region and the pull-up is in saturation, thus the pull-up and pull-down current must equate:

$$\beta_{PD}(V_{IN} - V_{TH} - \frac{V_{LOW}}{2})V_{LOW} = \frac{\beta_{PU}}{2}(V_{DEP})^2$$

Assuming that $V_{IN} = V_{DD}$ and ignoring $V_{LOW}/2$, we have

$$V_{LOW} = \frac{\beta_{PU}}{\beta_{PD}} \frac{V_{DEP}^2}{V_{DD} - V_{TH}}$$

$$= \frac{1}{2k} \frac{V_{DEP}^2}{V_{DD} - V_{TH}}$$

$$where \quad k = \frac{(L/W)_{PU}}{(L/W)_{PD}}$$

it is assumed that μ, ε and T_{OX} are the same for the pull-up and the pull-down. The quantity k is called the ratio of the inverter. The dimensions of the transistor can thus be chosen to ensure that V_{LOW} is less than V_{TH}. Increasing k will reduce the V_{LOW} voltage and thus making the gap between V_{LOW} and V_{TH} larger.

A.3 Exercises

A.1 A typical value of the oxide layer is 700 Å (70 nm). The permitivity of gate insulator is 3.εo (εo=8.85×10^{-14} farads/cm) and the electron mobility is 8×10^{-2} m^2/volt.sec. If the width and length of the gate is 5 x 10^{-4} cm, determine the following:

(a) the gate capacitance;
(b) the equation for I_{DS} for the linear region;
(c) the current I_{DS} when V_{GS} =5V, V_{TH} =1V and V_{DS} =1V;
(d) the equation for I_{DS} in the saturated region.

A.2 An NMOS transistor has a silicon dioxide thickness of 70 nm, a gate width of 3 μm and gate length of 1.5 μm. Assuming a supply voltage of 10 V and a threshold voltage of 2 V, determine the following:

(a) the gate capacitance, in pF;
(b) the equations which define the drain-source current;
(c) the drain-source saturation current.

Using graph paper plot the voltage-current characteristics for input gate-source voltage of 0 V to 10 V in steps of 1 V. The steps of drain-source voltage should be 2 V. The table provided should help to produce these results.

On a separate sheet of graph paper plot the drain-source current against the gate-source voltage for a drain-source voltage of 5 V. Determine the channel resistance and estimate the transfer ratio g_m ($\Delta I_{DS}/\Delta V_{GS}$) when the gate-source voltage is 5 V.

A.4 Program to determine gate capacitance

Program A.2 to determine the gate capacitance and beta factor for a FET is given next. Test run A.1 shows a run using the parameters in tutorial question A.1.

Program A.2

```
/*    Program to determine the gate capacitance        */
/*    and beta of a FET                                */

#define      PICO     1e-12
#define      MILLI    1e-3
#define      E_O      8.854e-12    /* Farads/ metre  */

#define      TRUE  1
#define      FALSE 0

#include <stdio.h>

float    calc_capacitance(float e_r,float w,float l,float th);
float    calc_beta(float e_r,float mu,float w,float l,float th);
void     get_parameters(float *w,float *l,float *tox);
void     display_calculated(float cap,float beta);
void     get_float(char msg[],float min,float max,float *val);


int      main(void)
{
float    e_r,mu,gate_width,gate_length,Tox,gate_cap,beta;


   e_r=3;   mu=8e-2; /* dielectric constant of 3   */
   get_parameters(&gate_width,&gate_length,&Tox);
   gate_cap=calc_capacitance(e_r,gate_width,gate_length,Tox);
   beta=calc_beta(e_r,mu,gate_width,gate_length,Tox);
   display_calculated(gate_cap,beta);
   return(0);
}

void     get_parameters(float *w,float *l,float *th)
{
   get_float("Enter gate width  (m)       >>",0,1e-2,w);
   get_float("Enter gate length (m)       >>",0,1e-2,l);
   get_float("Enter oxide thickness (m)   >>",0,1e-3,th);
}
float calc_capacitance(float e_r,float w,float l,float th)
{
   return(E_O*e_r*w*l/th);
}
```

```
float calc_beta(float e_r,float mu,float w,float l,float th)
{
   return(mu*e_r*E_O*w/(th*l));
}

void     display_calculated(float cap,float beta)
{
   printf("Capacitance is %8.4f pF\n",cap/PICO);
   printf("Beta is %e mA/V\n",beta/MILLI);
}

void     get_float(char msg[],float min,float max,float *val)
{
char     instr[BUFSIZ];
int      rtn,okay;
   /* get a value between min and max */
   do
   {
      printf("%s",msg);
      gets(instr);
      rtn=sscanf(instr,"%f",val);
      if ((rtn!=1) || (*val<min) || (*val>max))
      {
         okay=FALSE;
         printf("Invalid input <%s>\n",instr);
      }
      else okay=TRUE;

   } while (!okay);
}
```

Test run A.1

```
Enter gate width  (m)       >>5e-6
Enter gate length (m)       >>5e-6
Enter oxide thickness (m)   >>70e-9
Capacitance is   0.0095 pF
Beta is 3.03566e-02 mA/V
```

IC Fabrication

B.1 Introduction

This Appendix discusses the main IC fabrication processes.

B.2 NMOS fabrication

NMOS transistors are formed in a p-type substrate. The NMOS fabrication process requires involves nine main steps.

1. DEVELOP SILICON WAFER. Initially a single crystal of silicon of high purity is doped with a p-type material (such as Boron). Wafers are normally circular of between 75 and 150 mm in diameter.

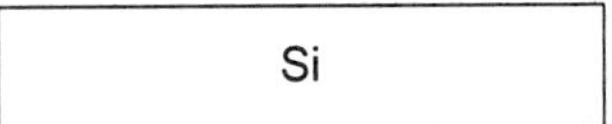

2. APPLY THIN LAYER OF SiO_2. A layer of silicon dioxide (SiO_2) is deposited on the surface of the wafer (approximately 1 μm thick). This protects the surface of silicon from damage and also provides an insulating substrate onto which other layers are deposited.

Oxide growth

Si

3. APPLY PHOTORESIST. Photoresist is applied to the surface and the wafer is spun to achieve an even distribution.

Protoresist

Si

4. APPLY MASK 1 ('THINOX'). The 'thinox' mask is applied and UV light

shines through the mask to harden the areas of photoresist. The 'thinox' mask defines the areas where transistor channels are formed.

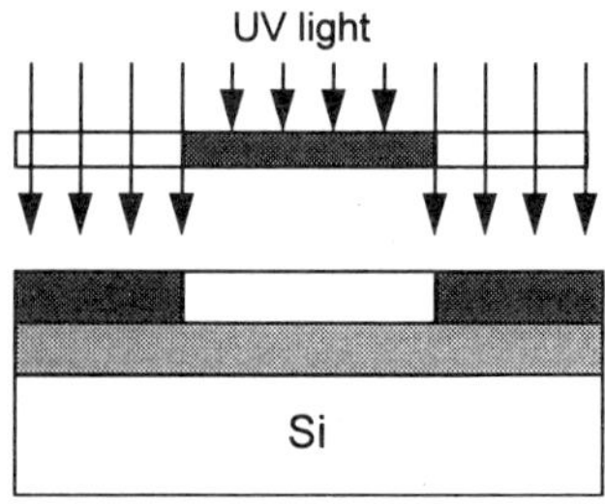

5. ETCH. Areas that have not been exposed are etched away.

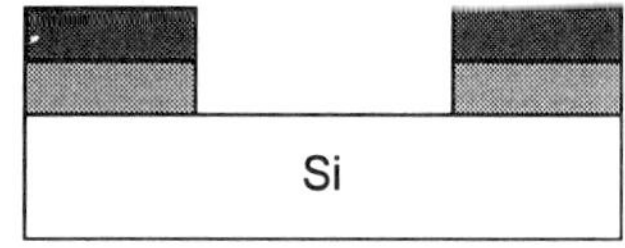

6. REMOVE PHOTORESIT. APPLY SiO_2. The remaining photoresist is removed and a thin layer of SiO_2 is deposited (approximately 0.1 μm).

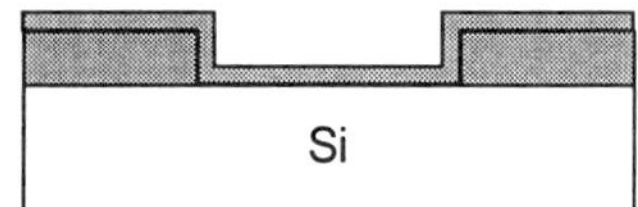

7. DEPOSIT POLYSILICON. APPLY PHOTORESIST. APPLY MASK 2 (POLYSILICON). ETCH. A thin layer of polysilicon is deposited (1-2 μm). Photoresist is again applied and the polysilicon is mask is applied. Again it is etched.

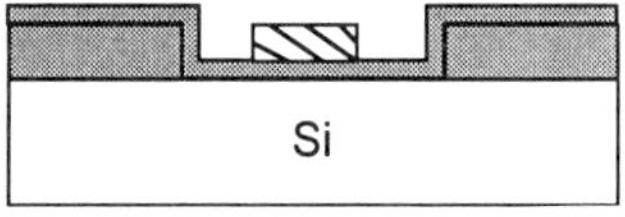

8. PHOTORESIST. APPLY MASK 3 (DIFFUSION). ETCH. DIFFUSE. Photoresist is deposited and the diffusion mask is applied and etched to expose areas where diffusion is to take place.

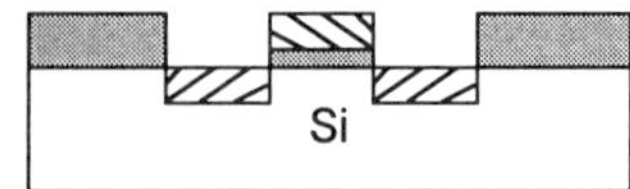

9. APPLY SiO_2. PHOTORESIST. APPLY MASK 4 (CONTACT CUTS). Apply SiO_2, photoresist and apply mask 4. Wafer is then etched to expose contact cuts.

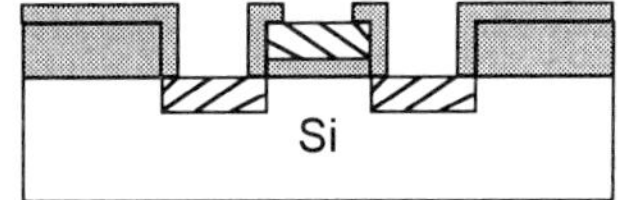

10. APPLY METAL. APPLY MASK 5 (METAL). ETCH. The whole chip then has metal (aluminium) deposited over its surface to a thickness of typically 1 µm. This metal layer is then masked and etched to form the required interconnection pattern.

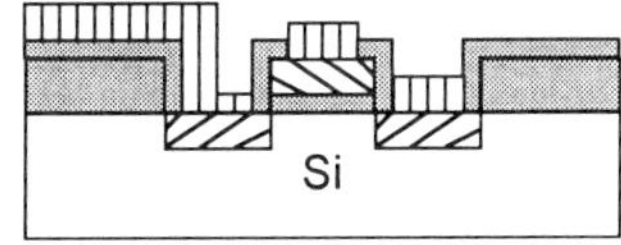

11. OVERGLASS. An overglass is applied to the wafer to protect the wafer from damage and contamination. Only the bonding pads are unglassed.

Notes:

1. When applying diffusion the wafer is heated to a high temperature and a gas containing the desired n-type impurity (such as phosphorus) is passed over the surface.
2. It normally takes a few hours to deposit the SiO_2 oxide in a wet-oxidation furnace at 1000°C and is grown on the silicon by heating either pure oxygen or in an atmosphere containing water vapour.
3. Photolithography is the method of patterning the layers of the IC.

B.3 CMOS fabrication

As discussed in the text there are several approaches to CMOS fabrication, including p-well, n-well, twin-tub and silicon-on-insulator.

B.3.1 p-well process

The p-well process uses an n-type substrate in which p-type transistors (PMOS) are fabricated. For n-type transistors (NMOS) a tub of p-type is diffused into the n-type substrate. The concentration and depth of the diffusion is important as it defines the threshold voltage of the transistor. A deep well or high well resistivity achieves a low threshold voltage. The main masks used are:

1. DEVELOP SILICON WAFER. APPLY SiO_2. APPLY PHOTORESIST. Initially a single crystal of silicon of high purity is doped with an n-type

material. SiO_2 and photoresist are applied to the surface.

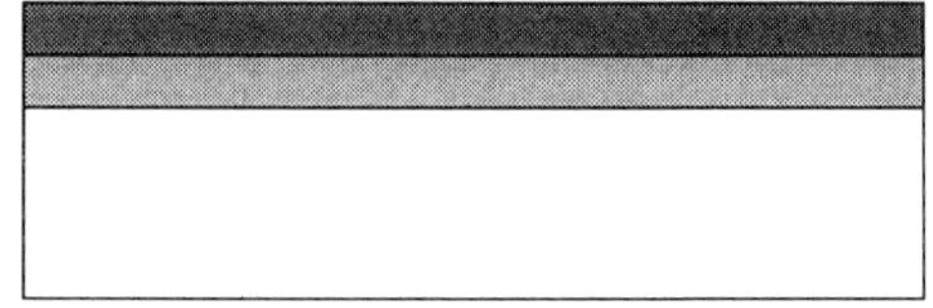

2. APPLY MASK 1 (P-WELL MASK). ETCH. DIFFUSE. Areas of p-well are defined and are diffused with p-type material.

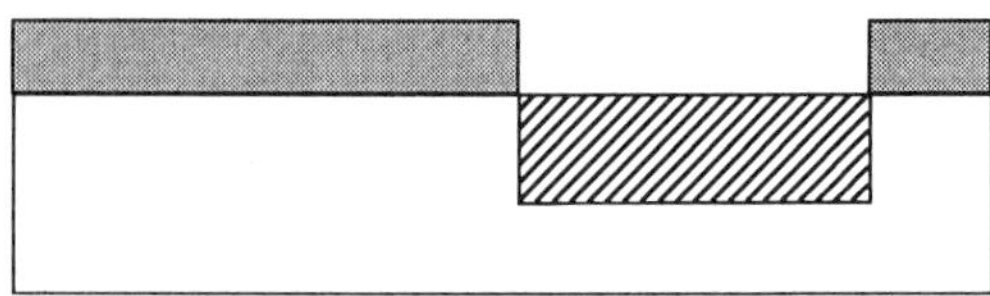

3. APPLY PHOTORESIST. APPLY MASK 2 ('THINOX'). ETCH. The thinox mask defines the areas where transistor channels are formed.

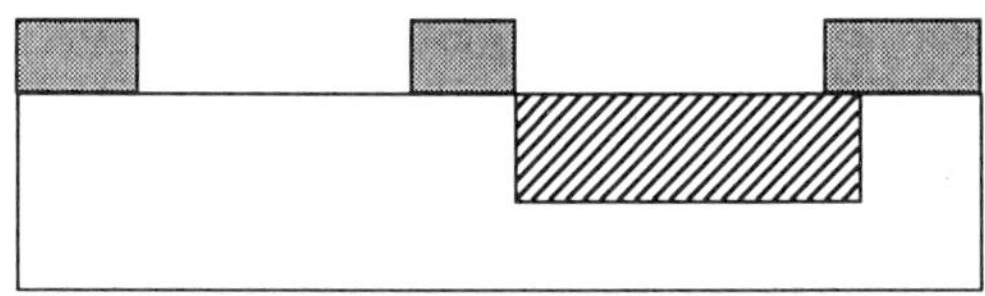

4. APPLY THIN LAYER OF SiO2. DEPOSIT POLYSILICON. APPLY PHOTORESIST. APPLY MASK 2 (POLYSILICON). ETCH. A thin layer of SiO2 and then a layer of polysilicon (1-2 μm). Photoresist is again applied and the polysilicon is mask is applied. The polysilicon is then etched to the required pattern.

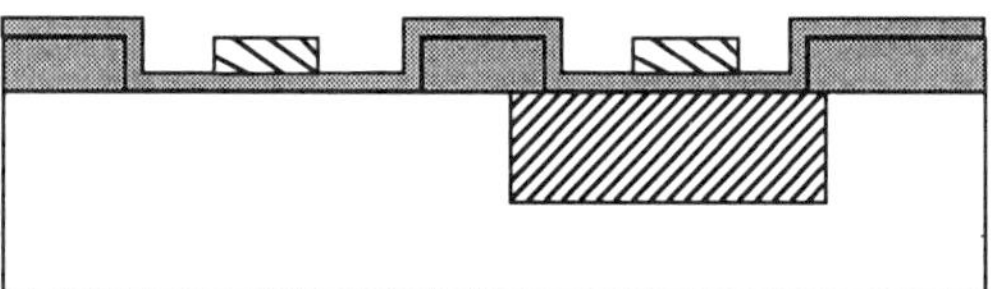

5. APPLY PHOTORESIST. APPLY MASK 4 ('THINOX'). ETCH. Photoresist is again applied and the thinox mask is mask is applied. Again it is etched.

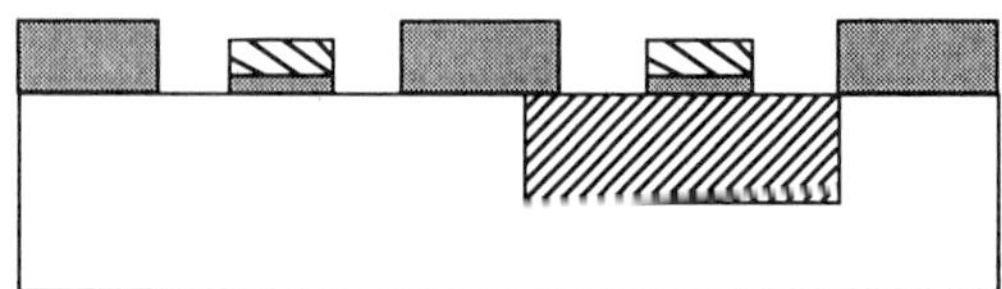

6. APPLY MASK 5 (POSITIVE OF P+ MASK). DIFFUSE WITH P-TYPE. The p+ mask and mask 2 are used to defined the areas in which n-type material is diffused in the substrate.

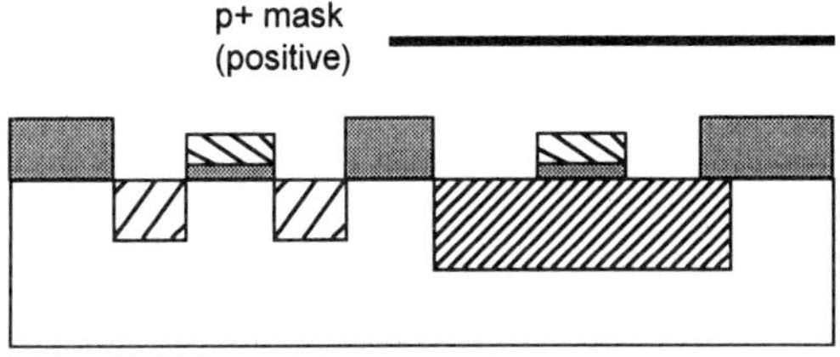

7. APPLY MASK 6 (NEGATIVE OF P+ MASK). DIFFUSE WITH N-TYPE. A negative of the p+ mask and mask 2 are used to define the areas in which n-type material is diffused in the substrate.

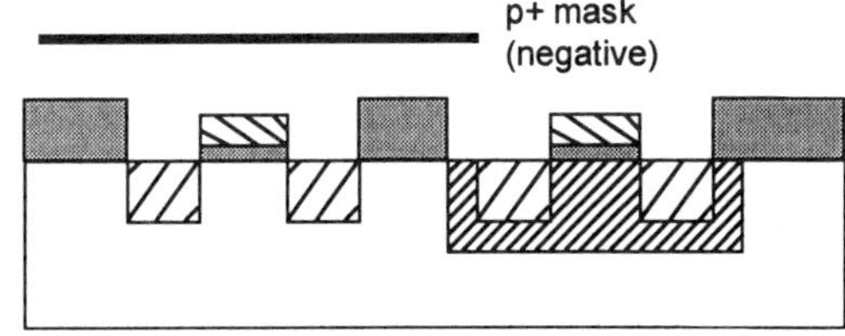

9. APPLY SiO_2. PHOTORESIST. APPLY MASK 7 (CONTACT CUTS). Apply SiO_2, photoresist and apply mask 4. Wafer is then etched to expose contact cuts.

10. APPLY METAL. APPLY MASK 8 (METAL). ETCH. The whole chip then has metal (aluminium) deposited over its surface to a thickness of typically 1 μm. This metal layer is then masked and etched to form the required interconnection pattern.

11. OVERGLASS. An overglass is applied to the wafer to protect the wafer from damage and contamination. Only the bonding pads are unglassed.

CMOS/NMOS DESIGN RULES

C.1 Introduction

This Appendix outlines typical process specification for CMOS.

C.2 1.2 micron single polysilicon double metal n-well and p-well design rules and process specifications

The micron processes do not rely on lambda-based rules and define minimum and maximum distances by absolute distances. Table C.1 outlines the minimum thickness of the oxide and Table C.2 defines the minimum separation distance. Table C.3 gives typical capacitance values.

Table C.1 Minimum thickness.

	Minimum (nm)
Gate oxide	22.5
Field oxide	600
Polysilicon	400
Intermediate oxide	600
Metal 1	600
Metal 2	1150

Table C.2 Minimum separation (nm).

	Minimum (nm)
Metal 1 to polysilicon	600
Metal 1 to substrate	1100
Metal 1 to diffusion	600
Metal 2 to polysilicon 1	1300
Metal 2 to substrate	2000
Metal 2 to diffusion	1300
Metal 2 to metal 1	700

Table C.3 Typical capacitance (10^{-4}pF/μm^2).

	Typical (10^{-4}pF/μm^2)
Metal 1 to polysilicon	0.6
Metal 1 to substrate	0.33
Metal 1 to diffusion	0.6
Metal 2 to polysilicon 1	0.28
Metal 2 to substrate	0.17
Metal 2 to diffusion	0.28
Metal 2 to metal 1	0.50

C.3 2.0 micron double polysilicon double metal n-well CMOS process specification

The minimum and maximum thickness are of the oxide are given in Table C.4 and the minimum and maximum thickness of conductors is given in Table C.5. Sheet resistance values are given in Table C.5.

Table C.4 Minimum and maximum thickness.

Oxide thickness	*Minimum*	*Typical*	*Maximum*
Polysilicon 1	370	400	430
Polysilicon 2	470	500	530
Field oxide	5500	6000	6500
Metal 1 to polysilicon	8000	8500	9000
Metal 1 to substrate	13 500	14 500	15 500
Metal 1 to n+/p+ diffusion	8500	9000	9500
Metal 2 to metal 1	6000	6500	7500
Polysilicon 1 to polysilicon 2	650	750	850

Table C.5 Minimum and maximum conductor thickness.

Conductor thickness	*Minimum*	*Typical*	*Maximum*
Polysilicon 1	3700	4000	4300
Polysilicon 2	3700	4000	4300
Metal 1	5500	6000	6500
Metal 2	10 500	11 500	12 500

Table C.6 Minimum and maximum of sheet resistance (Ω per square).

Sheet resistance (Ω per square)	*Minimum*	*Typical*	*Maximum*
P+ active	40	57	80
N+ active	20	28	40
Polysilicon 1	15	21	30
Polysilicon 2	18	25	30
Metal 1	0.050	0.070	0.090
Metal 2	0.030	0.040	0.050
N-well	2000	2500	3000

C.4 CIF format

Caltech. Intermediate Form (CIF) code is a low-level graphics language to define the geometry of integrated circuits. It is a standard format text file which most design packages and fabrication plants can read. It uses an x,y coordinate system.

C.4.1 Geometric shapes

The three main geometries are boxes (`B`), polygons (`P`) and wires (`W`). A box (`B`) of length (`L`), width (`W`), centre coordinate (`Cx, Cy`) and direction (`D`) is defined as:

```
B  L  W  Cx,Cy D
```

A polygon with coordinates (`x1`,`y1`), (`x2`,`y2`), ... is defined as:

```
P  x1 y1 x2 y2 ...
```

The connection between the first and last points is assumed. A wire of width (`Width`) with coordinates (`x1`,`y1`), (`x2`,`y2`), ... is defined as:

```
W Width  x1 y1 x2 y2 ...
```

Figure C.1 shows examples of the three formats. The wire has a width of 10 and uses the coordinates (15,60), (15,95) and (60,95). The box has a length of 50, a width of 30 and has a centre point at (75,45). The polygon in the example is connected by the points (20,-10), (10,50) and (40,10).

C.4.2 Layers

The current layer for the geometric shape is defined by a layer definition. Table C.7 defines these layers and their names.

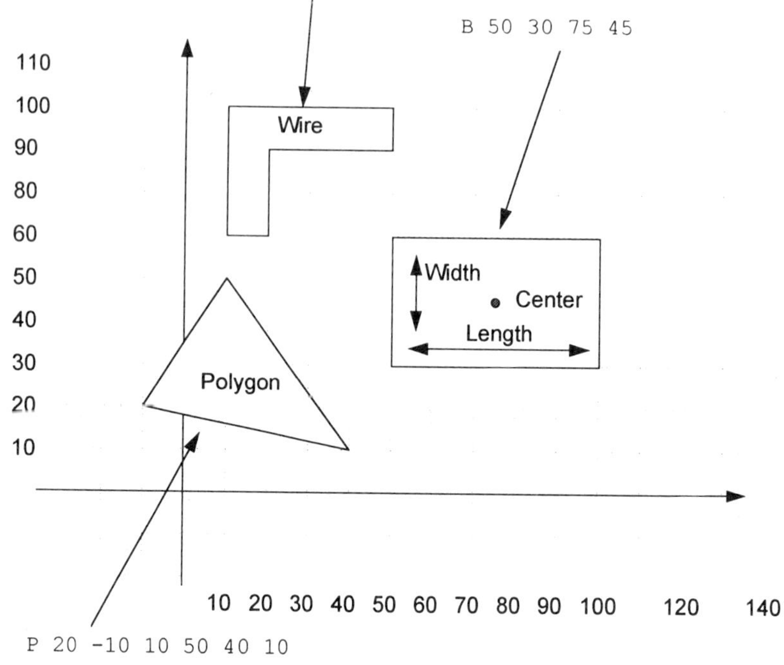

Figure C.1 Example geometric shapes.

Table C.7 CIF layer names.

CMOS layer	*Name*	*NMOS layer*	*Name*
CMOS diff/thinox	CAA	NMOS diff/thinox	ND
CMOS nDiff/thinox	CNA		
CMOS pDiff/thinox	CPA	NMOS pDiff/thinox	
CMOS polysilicon 1	CPF	NMOS polysilicon	NP
CMOS polysilicon 2	CPS		
CMOS Contact cut	CC	NMOS Contact cut	NC
CMOS metal 1	CMF	NMOS metal 1	NM
CMOS metal 2	CMS	NMOS metal 2	NN
		NMOS implant	NI
CMOS p+ mask	CS or CPP		
CMOS p-well	CW or CPW		
CMOS via	CVA	NMOS via	NV
CMOS via	CVA	NMOS buried contact	NB
CMOS via	CVA	NMOS via	NV

The current layer is defined with the `L` keyword. An example excerpt from a CIF file is given next:

```
L NM;
      W 500 0,400 2500,400;
      W 500 0,300 2500,3000;
L ND;
      B 500 500 500,250;
L NC;
      B 250 250 500,250;
```

TRAFFIC LIGHT CONTROLLER SIMULATOR D

D.1 Introduction

This Appendix contains a program which simulates the traffic light controller. The equations used are:

$$RED\_EW = G$$

$$RED\_NS = \overline{G}$$

$$YELLOW\_EW = \overline{G}.F.E$$

$$YELLOW\_NS = G.F.E$$

$$GREEN\_EW = \overline{G}.\overline{F} + \overline{G}.\overline{E}$$

$$GREEN\_NS = G.\overline{F} + G.\overline{E}$$

$$WALK\_EW = \overline{F}.\overline{G}$$

$$DONT\_EW = G + A.F.\overline{E} + \overline{G}.F.E$$

$$WALK\_NS = G.\overline{F}$$

$$DONT\_NS = \overline{G} + A.F.\overline{E} + G.F.E$$

D.2 Program and test results

Program D.1

```
#include <stdio.h>

#define  TRUE  1
#define  FALSE 0

int   not(int x);
int   or2(int x,int y);
int   or3(int x,int y,int z);
int   and2(int x,int y);
int   and3(int x,int y,int z);
```

```
int   main(void)
{
int   a,b,c,d,e,f,g;
int   seq,red_ns,red_ew,yellow_ns,yellow_ew,green_ns,green_ew;
int   walk_ns,dont_ns,walk_ew,dont_ew;

    printf("E-W   N-S   E-W N-S\n");
   printf("R A G R A G W D W D\n");

   for (seq=1;seq<128;seq++)
   {
      a=(seq&1); b=(seq&2)>>1; c=(seq&4)>>2; d=(seq&8)>>3;
      e=(seq&16)>>4; f=(seq&32)>>5; g=(seq&64)>>6;
      red_ew=g;
      red_ns=not(g);
      yellow_ew=and3(not(g),f,e);
      yellow_ns=and3(g,f,e);
      green_ew=or2(and2(not(g),not(f)),and2(not(g),not(e)));
      green_ns=or2(and2(g,not(f)),and2(g,not(e)));
      walk_ew=and2(not(f),not(g));
      dont_ew=or3(g,and3(a,f,not(e)),and3(not(g),f,e));
      walk_ns=and2(not(f),g);
      dont_ns=or3(not(g),and3(a,f,not(e)),and3(not(g),f,e));

      if ((seq % 20)==0)
         { printf("Press any key to continue >>"); getchar(); }
      printf("%d %d %d %d %d %d %d %d %d %d\n",
      red_ew,yellow_ew,green_ew,
      red_ns,yellow_ns,green_ns,
      walk_ew,dont_ew,walk_ns,dont_ns);

   }
   return(0);
}

int   and2(int x,int y)
{
   return(x && y);
}

int   and3(int x,int y,int z)
{
   return(x && y && z);
}

int   or2(int x, int y)
{
   return((x || y));
}

int   or3(int x, int y, int z)
{
   return((x || y || z));
}

int   not(int x)
{
   return(!x);
}
```

Test run D.1

```
E-W    N-S    E-W N-S
R A G R A G W D W D
0 0 1 1 0 0 1 0 0 1
0 0 1 1 0 0 1 0 0 1
0 0 1 1 0 0 1 0 0 1
0 0 1 1 0 0 1 0 0 1
0 0 1 1 0 0 1 0 0 1
0 0 1 1 0 0 1 0 0 1
0 0 1 1 0 0 1 0 0 1
0 0 1 1 0 0 1 0 0 1
0 0 1 1 0 0 1 0 0 1
0 0 1 1 0 0 1 0 0 1
0 0 1 1 0 0 1 0 0 1
0 0 1 1 0 0 1 0 0 1
0 0 1 1 0 0 1 0 0 1
0 0 1 1 0 0 1 0 0 1
0 0 1 1 0 0 1 0 0 1
0 0 1 1 0 0 1 0 0 1
0 0 1 1 0 0 1 0 0 1
0 0 1 1 0 0 1 0 0 1
0 0 1 1 0 0 1 0 0 1
Press any key to continue >>
0 0 1 1 0 0 1 0 0 1
0 0 1 1 0 0 1 0 0 1
0 0 1 1 0 0 1 0 0 1
0 0 1 1 0 0 1 0 0 1
0 0 1 1 0 0 1 0 0 1
0 0 1 1 0 0 1 0 0 1
0 0 1 1 0 0 1 0 0 1
0 0 1 1 0 0 1 0 0 1
0 0 1 1 0 0 1 0 0 1
0 0 1 1 0 0 1 0 0 1
0 0 1 1 0 0 1 0 0 1
0 0 1 1 0 0 1 0 0 1
0 0 1 1 0 0 0 0 0 1
0 0 1 1 0 0 0 1 0 1
0 0 1 1 0 0 0 0 0 1
0 0 1 1 0 0 0 1 0 1
0 0 1 1 0 0 0 0 0 1
0 0 1 1 0 0 0 1 0 1
0 0 1 1 0 0 0 0 0 1
0 0 1 1 0 0 0 1 0 1
Press any key to continue >>
0 0 1 1 0 0 0 0 0 1
0 0 1 1 0 0 0 1 0 1
0 0 1 1 0 0 0 0 0 1
0 0 1 1 0 0 0 1 0 1
```

```
0 0 1 1 0 0 0 0 0 1
0 0 1 1 0 0 0 1 0 1
0 0 1 1 0 0 0 0 0 1
0 0 1 1 0 0 0 1 0 1
0 1 0 1 0 0 0 1 0 1
0 1 0 1 0 0 0 1 0 1
0 1 0 1 0 0 0 1 0 1
0 1 0 1 0 0 0 1 0 1
0 1 0 1 0 0 0 1 0 1
0 1 0 1 0 0 0 1 0 1
0 1 0 1 0 0 0 1 0 1
0 1 0 1 0 0 0 1 0 1
0 1 0 1 0 0 0 1 0 1
0 1 0 1 0 0 0 1 0 1
0 1 0 1 0 0 0 1 0 1
0 1 0 1 0 0 0 1 0 1
Press any key to continue >>
0 1 0 1 0 0 0 1 0 1
0 1 0 1 0 0 0 1 0 1
0 1 0 1 0 0 0 1 0 1
0 1 0 1 0 0 0 1 0 1
1 0 0 0 0 1 0 1 1 0
1 0 0 0 0 1 0 1 1 0
1 0 0 0 0 1 0 1 1 0
1 0 0 0 0 1 0 1 1 0
1 0 0 0 0 1 0 1 1 0
1 0 0 0 0 1 0 1 1 0
1 0 0 0 0 1 0 1 1 0
1 0 0 0 0 1 0 1 1 0
1 0 0 0 0 1 0 1 1 0
1 0 0 0 0 1 0 1 1 0
1 0 0 0 0 1 0 1 1 0
1 0 0 0 0 1 0 1 1 0
1 0 0 0 0 1 0 1 1 0
1 0 0 0 0 1 0 1 1 0
1 0 0 0 0 1 0 1 1 0
1 0 0 0 0 1 0 1 1 0
Press any key to continue >>
1 0 0 0 0 1 0 1 1 0
1 0 0 0 0 1 0 1 1 0
1 0 0 0 0 1 0 1 1 0
1 0 0 0 0 1 0 1 1 0
1 0 0 0 0 1 0 1 1 0
1 0 0 0 0 1 0 1 1 0
1 0 0 0 0 1 0 1 1 0
1 0 0 0 0 1 0 1 1 0
1 0 0 0 0 1 0 1 1 0
1 0 0 0 0 1 0 1 1 0
1 0 0 0 0 1 0 1 1 0
```

```
1 0 0 0 0 1 0 1 1 0
1 0 0 0 0 1 0 1 1 0
1 0 0 0 0 1 0 1 1 0
1 0 0 0 0 1 0 1 1 0
1 0 0 0 0 1 0 1 1 0
1 0 0 0 0 1 0 1 0 0
1 0 0 0 0 1 0 1 0 1
1 0 0 0 0 1 0 1 0 0
1 0 0 0 0 1 0 1 0 1
Press any key to continue >>
1 0 0 0 0 1 0 1 0 0
1 0 0 0 0 1 0 1 0 1
1 0 0 0 0 1 0 1 0 0
1 0 0 0 0 1 0 1 0 1
1 0 0 0 0 1 0 1 0 0
1 0 0 0 0 1 0 1 0 1
1 0 0 0 0 1 0 1 0 0
1 0 0 0 0 1 0 1 0 1
1 0 0 0 0 1 0 1 0 0
1 0 0 0 0 1 0 1 0 1
1 0 0 0 0 1 0 1 0 0
1 0 0 0 0 1 0 1 0 1
1 0 0 0 1 0 0 1 0 0
1 0 0 0 1 0 0 1 0 0
1 0 0 0 1 0 0 1 0 0
1 0 0 0 1 0 0 1 0 0
1 0 0 0 1 0 0 1 0 0
1 0 0 0 1 0 0 1 0 0
1 0 0 0 1 0 0 1 0 0
1 0 0 0 1 0 0 1 0 0
Press any key to continue >>
1 0 0 0 1 0 0 1 0 0
1 0 0 0 1 0 0 1 0 0
1 0 0 0 1 0 0 1 0 0
1 0 0 0 1 0 0 1 0 0
1 0 0 0 1 0 0 1 0 0
1 0 0 0 1 0 0 1 0 0
1 0 0 0 1 0 0 1 0 0
1 0 0 0 1 0 0 1 0 0
```

VHDL REFERENCE

E.1 Conventions

The conventions used in this reference are:

1. Reserved words are written in `typeface`
2. Curly brackets [{ }] denote optional items.

E.2 Syntax

Statement	*Syntax*	*Example*

Architecture	architecture name is
	{ signal declarations }
	begin
	{ concurrent statements }
	end architecture
		architecture SEQ_SIGNALS **of** EXAMPLES **is**
		begin
		process(sig1);
		begin
		sig2<=sig1;
		z<=sig2;
		end process;
		end architecture;

Case	case boolean_statement is when choices => { sequential statement(s) } when choices => { sequential statement(s) } { ... } end case;	**case** SELECT **is** **when** 0 => Z=D0; **when** 1 => Z=D1; **when** 2 => Z=D2; **when** 3 => Z=D3; **end case**
Entity	entity entity_name is { port declaration(s) } end entity_name;	
Exit		**exit**; **exit when** x<y; **exit** loop_label **when** x'
For	for identifier in discrete_range loop { sequential statements } end loop;	**for** i **in** 0 **to** 255 **loop** x<=i; **wait** 100 ms; **end loop**;

If	`if` boolean_statement `then`
	{ sequential statement(s) }
	`elsif` boolean_statement `then`
	{ sequential statement(s) }
	`else`
	{ sequential statement(s) }
	`end if;`
		if set = '1' **and** reset = '0' **then**
		q<='0' **after** 10 ns;
		qbar<='1' **after** 15 ns;
		elsif set ='0' **and** reset='1' **then**
		q<='1' **after** 15 ns;
		qbar<='0' **after** 10 ns;
		elsif set ='1' **and** reset='1' **then**
		q<='1' **after** 15 ns;
		qbar<='1' **after** 15 ns;
		endif

Loop	`loop`
	{sequential statement(s) }
	`end loop;`
		loop
		wait until clock='1';
		count_val:=(count_val-1);
		if (count_val=-1) **then**
		count_val:=15;
		count<=count_val;
		end loop;

Process	process_name:
	`process`(sensitivity_list)
	`begin`
	{ sequential_statement(s) }
	`end` process_name;
		process(sig1);
		begin
		sig2<=sig1;
		z<=sig2;
		end process;

Wait	`wait on` sensitivity list; `wait until` boolean-expression; `wait for` time-expression;	**wait on** sig1, sig2, sig3; **wait for** 50 ns; **wait until** (sig1/=sig2);

PSpice Tutorial

F.1 Introduction

This appendix contains a practical Spice simulation using PSpice version 6.0 for WINDOWS. Figure F.1 shows the directory listing of the PSpice program. Included in this directory are the programs:

- Psched – is a schematic capture package;
- Probe – plots graphs of simulations;
- PSpice – performs the PSpice simulation
- Stmed – stimulus editor.

Figure F.1 PSpice programs.

F.2 Simulation process

The basic simulation is shown in Figure F.2. Initially Psched is used to create a schematic. The output file from Psched is *file*.cir. This contains the PSpice information required for the PSpice simulation. The files produced by the simulator are a text file (*file*.out) and a graph plotting data file (*file*.dat).

Each of the programs can be run individually but the schematic capture package can then run automatically. Figure F.3 shows that the menu option Analysis contains the options Simulate and Run Probe (to run Probe). These can also be run using the functions keys F11 and F12.

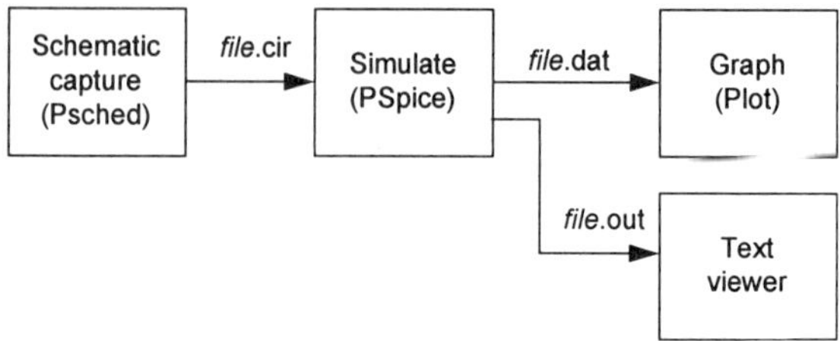

Figure F.2 Design process.

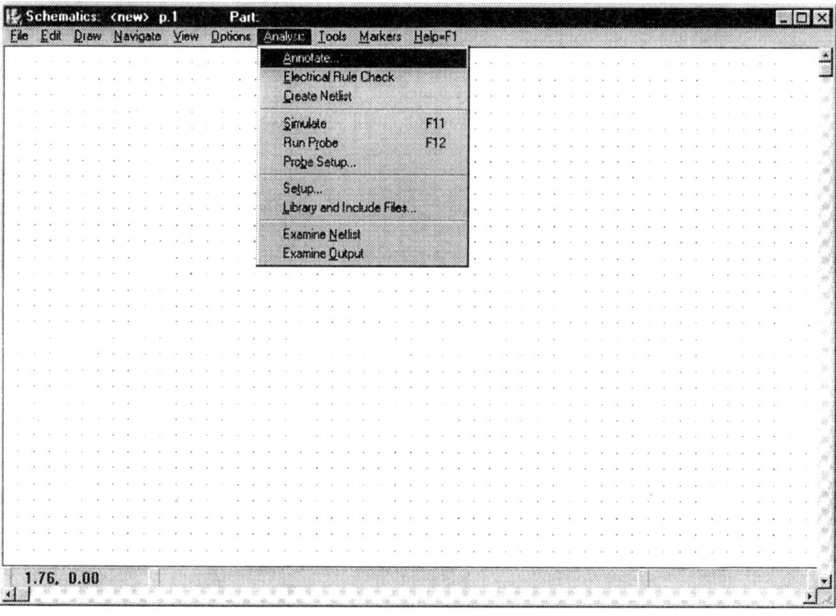

Figure F.3 Running other applications for Psched.

F.3 Schematic capture

The Psched (PSpice schematic edit) package generates the schematic. Figure F.4 shows the initial screen. The Draw menu option from the main menu bar is used to get components (Cntrl-G), draw wires (Ctrl-W) and busses (Cntrl-B), place text (Cntrl-T), place a part (Cntrl-P) and to repeat the last operation (Space). This menu option is shown in Figure F.5. The hot keys allow the draw options to be accessed quickly. For example a wire can be added by simply pressing the Cntrl and W keys (at the same time).

Figure F.4 Design process.

File Edit Draw Navigate View Options

Repeat	Space
Place Part	Ctrl+P
Text...	Ctrl+T
Wire	Ctrl+W
Bus	Ctrl+B
Block	
Get New Part...	Ctrl+G
Rewire	Ctrl+D

Figure F.5 Draw menu option.

To initially get a part the Get New Part option in the Draw menu is selected. Figure F.6 shows the entry of a resistor part. If the part is unknown the Browse... menu button can be selected.

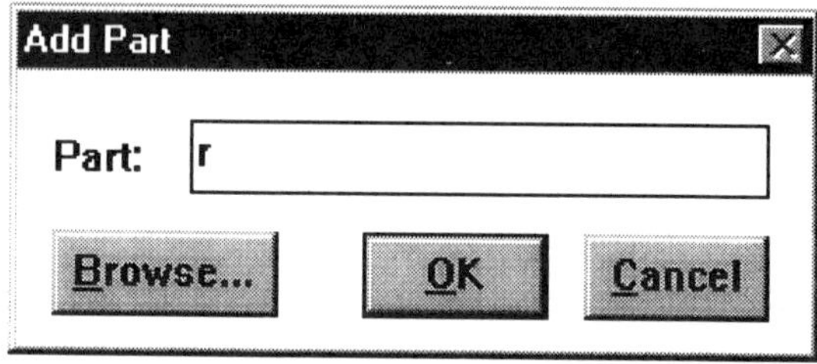

Figure F.6 Getting a resistor part.

Figure F.7 shows the placement of two resistors. (Note that the package automatically labels them as R1 and R2). To rotate R2 it is first selected by a single click on the resistor. Then either Cntrl-F is pressed or the Rotate option from the Edit menu is selected (as shown in Figure F.8). The updated schematic with R2 rotated is also shown in Figure F.8.

To edit a part (such as to change its name, its values, and so on), the mouse cursor is double-clicked twice on the part. Figure F.9 shows the edit screen for the resistor R1. In this case the value of the resistor has been changed to 10 kΩ. Note that the value of the resistance can also be changed by double clicking on the resistor value on the schematic, the window which appears is shown in Figure F.10. Figure F.11 shows the schematic after the value of the resistors has been changed to 10 kΩ. A capacitor (C1) has also been added. It's part name entered is C (for capacitor) and the schematic shows that its value has been changed from 1 nF to 100 nF.

The labels on a part (such as R1 and 10k) can be changed by pressing the left mouse key down and dragging the label until the correct position is reached.

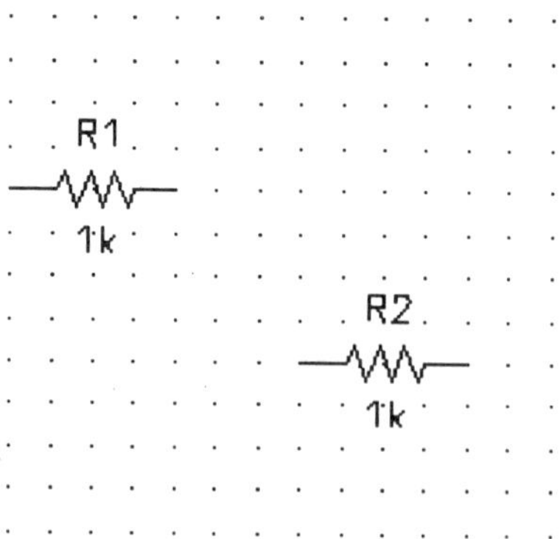

Figure F.7 Placement of resistors.

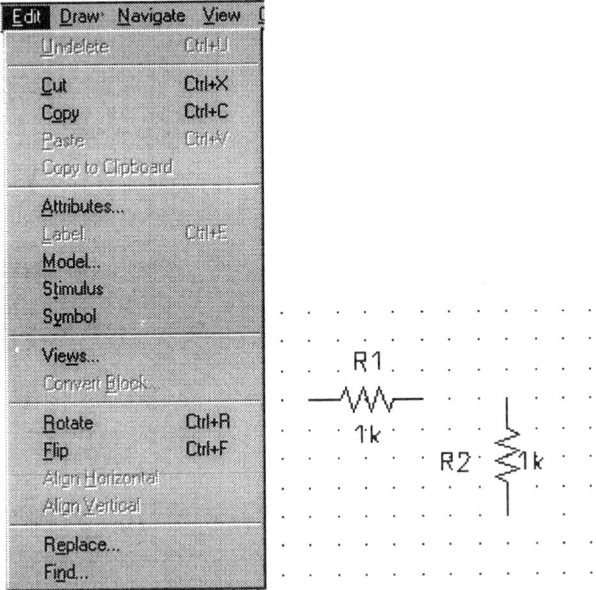

Figure F.8 Edit menu.

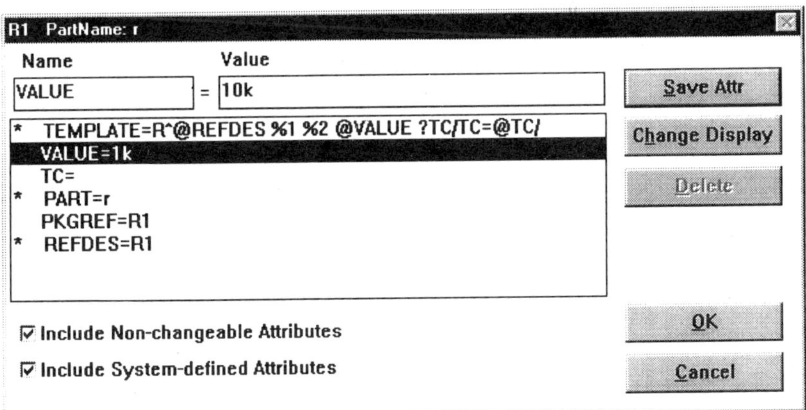

Figure F.9 Edit part window.

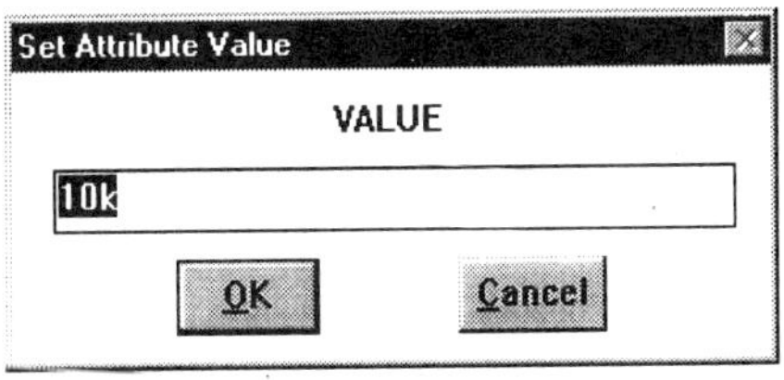

Set Attribute Value
VALUE
10k
OK
Cancel

Figure F.10 Setting resistor value.

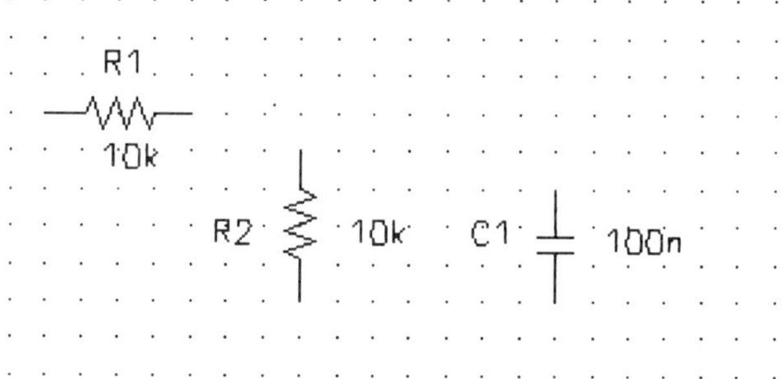

Figure F.11 Updated schematic with two resistors and a capacitor.

Next, wires are added to the schematic by initially pressing Cntrl-W (or selecting Wire option for the Draw menu). A pencil cursor should then be shown and the start of the wire is selected with the first click of the mouse button. Then the next mouse button click selects the end of the wire, as shown in Figure F.12.

The completed schematic with a voltage source (V1) and an analogue ground connection is shown in Figure F.13. An analogue ground is added by selecting Browse... from the Get Part menu window. Figure F.14 shows the window that appears after this. The library is set to port.slb and the AGND part is then selected (as shown). A simple voltage source is selected in a similar way and is the Get Part windows is shown in Figure F.15. It is contained in the source.slb library. Next the AC voltage for the input voltage is added. Figure F.16 shows how the AC characteristic is set to 10 V.

A label, Vout, has also been added by selecting the required wire (or net) and adding a label.

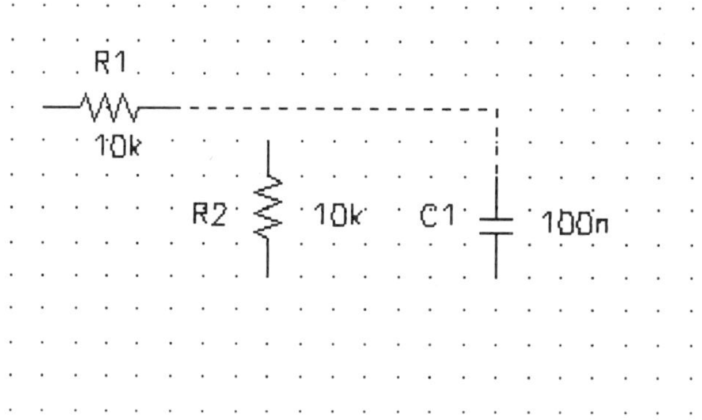

Figure F.12 Drawing a wire.

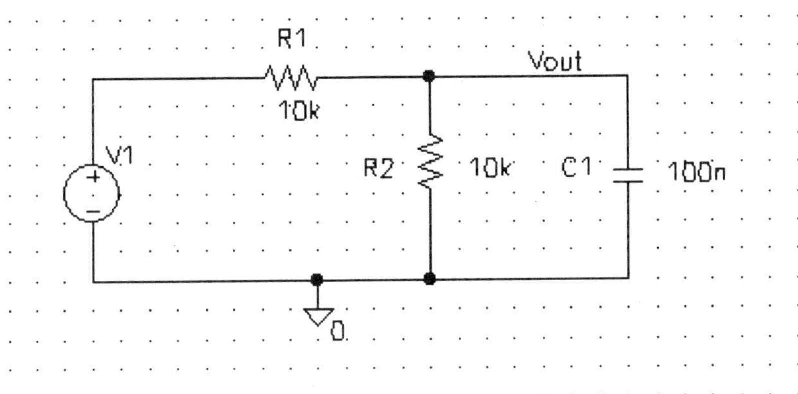

Figure F.13 Updated schematic.

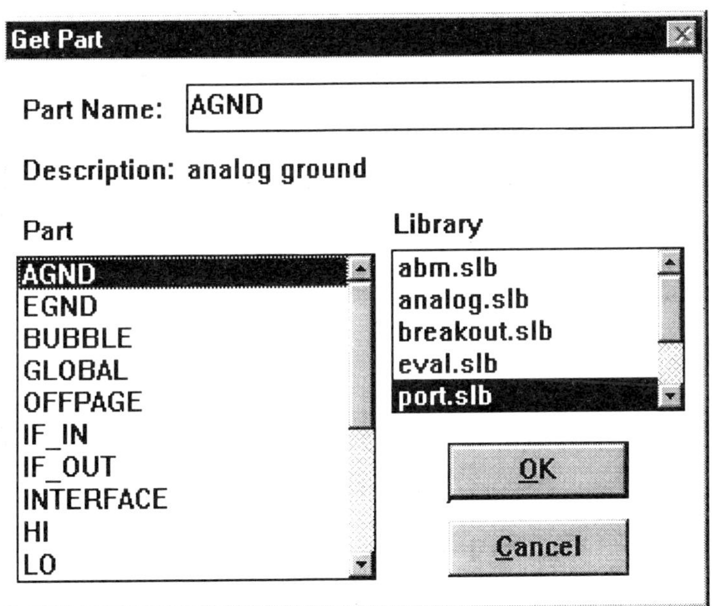

Figure F.14 Analogue ground part.

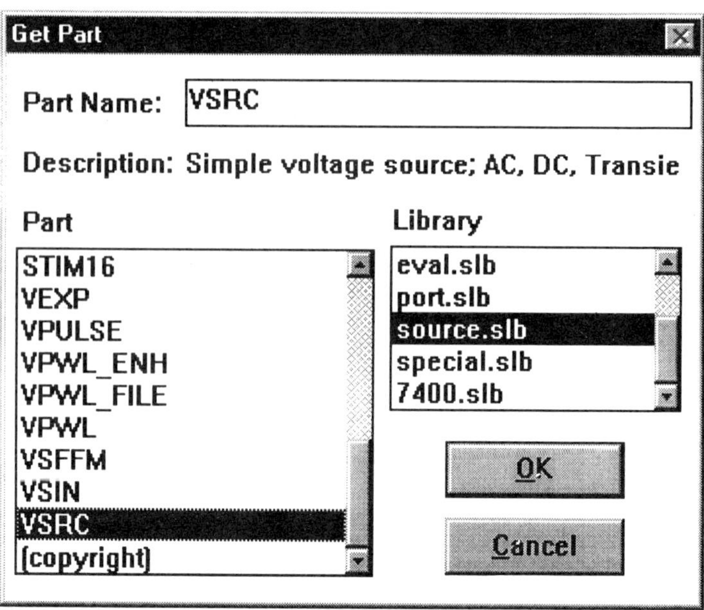

Figure F.15 Voltage source part.

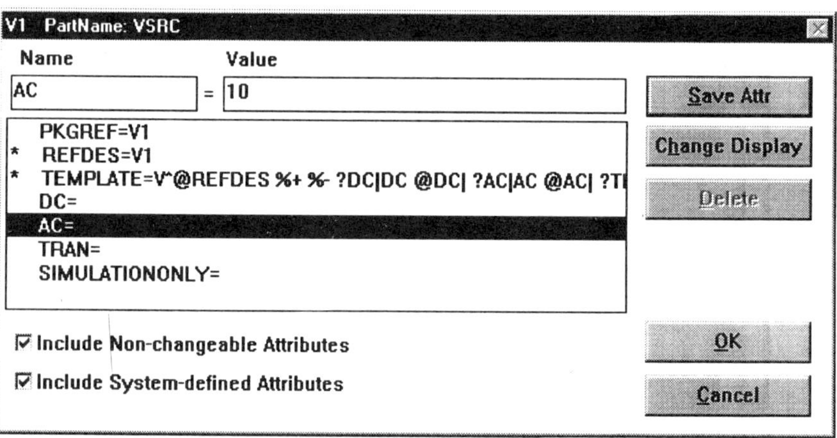

Figure F.16 Setting voltage source.

F.4 Saving file

The schematic can now be saved with the File→Save (or File→Save As) menu options. In this case save the file as New. This will save the schematic as New.sch.

F.5 Electrical checking

The schematic must now be checked for an errors, such as dangling wires, incorrectly connected components, and so on. To achieve this the

Analysis→Electrical Check

option is selected. Figure F.17 shows an example error. In this case the pin 1 of R1 is not connected to anything (that is, it is floating). One of the pins of V1 is also floating. All errors produced by the Electrical Check must be fixed before a simulation can start.

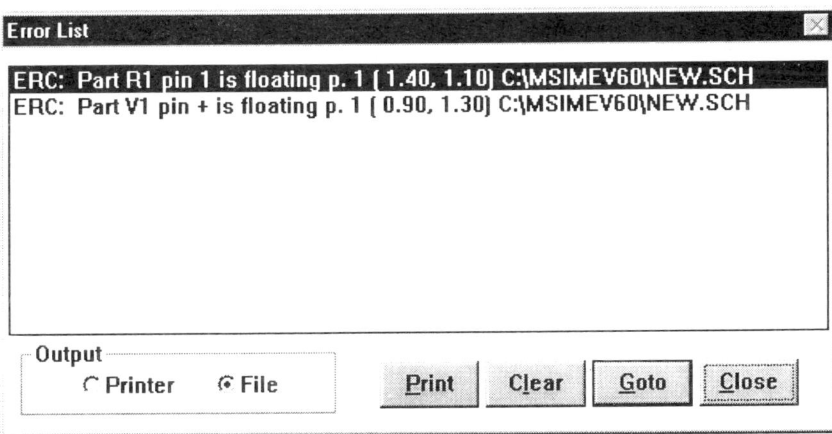

Figure F.17 Errors in electrical check.

Once the electrical checking is complete then the status line at the bottom of the main schematic windows should say ERC complete.

F.6 Specifying an analysis

Next the type of analysis to be conducted can be specified. To achieve this the

Analysis→Setup

option is selected and the windows displayed should be similar to the one shown in Figure F.18. The analysis types are AC sweep, DC sweep, Bias Point Analysis, Transfer Function and Transient Analysis. Figure F.19 shows how the AC frequency sweep is specified. In this case the frequency sweep ranges from 10 Hz to 100 kHz with 100 steps in frequency. Note that the enabled box at the side of the analysis type in the Analysis Setup window must be checked (ticked) for it to be valid.

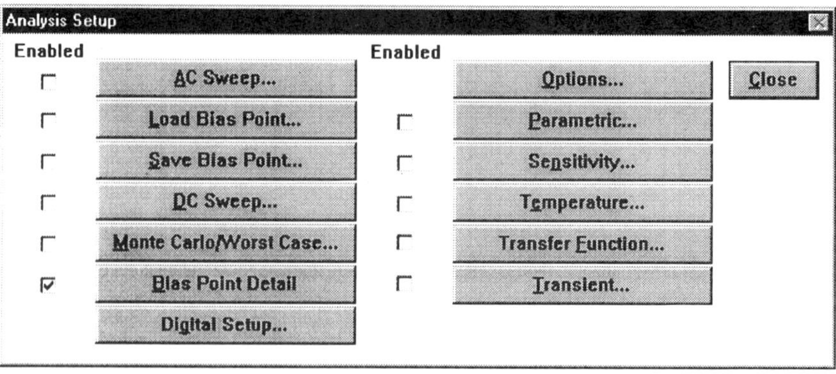

Figure F.18 Analysis window.

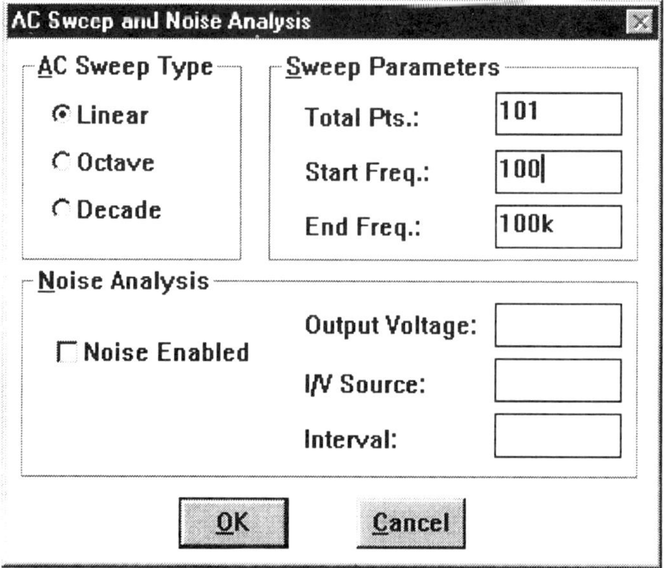

Figure F.19 AC sweep parameters.

F.7 Netlist

After the file has been saved there are three files which are used with PSpice, these are NEW.SCH (main schematic file), NEW.ALS (alias file) and NEW.NET (netlist file). The contents of these is given in Listing F.1 to F.3. Notice that the simulation data (`.ac LIN 101 100 100k`) has been automatically put into the NEW.SCH file.

Listing F.1: CONTENTS OF NEW.SCH

```
* C:\MSIMEV60\NEW.SCH

* Schematics Version 6.0 - January 1994
* Sat Sep 21 21:08:58 1996

** Analysis setup **
.ac LIN 101 100 100k
.OP

* From [SCHEMATICS NETLIST] section of msim.ini:
.lib nom.lib

.INC "NEW.net"
.INC "NEW.als"

.probe

.END
```

Listing F.2: CONTENTS OF NEW.ALS

```
* Schematics Aliases *

.ALIASES
R_R1          R1(1=$N_0001 2=Vout )
R_R2          R2(1=0 2=Vout )
C_C1          C1(1=0 2=Vout )
V_V1          V1(+=$N_0001 -=0 )
_    _(Vout=Vout)
.ENDALIASES
```

Listing F.3: CONTENTS OF NEW.NET

```
* Schematics Netlist *

R_R1       $N_0001 Vout 10k
R_R2       0 Vout 10k
C_C1       0 Vout 100n
V_V1       $N_0001 0  AC 10
```

F.8 Simulation

Next a simulation takes place with PSpice. To achieve this the

Analysis→Simulate

menu option is selected. Figure F.20 shows a successful simulation. An errors in the simulation will not mean that no output data is produced (*file*.out). In this case the output file is NEW.OUT. Normally after a successful simulation the Probe package is automatically called (the .probe line in the schematic file causes Probe to be run automatically).

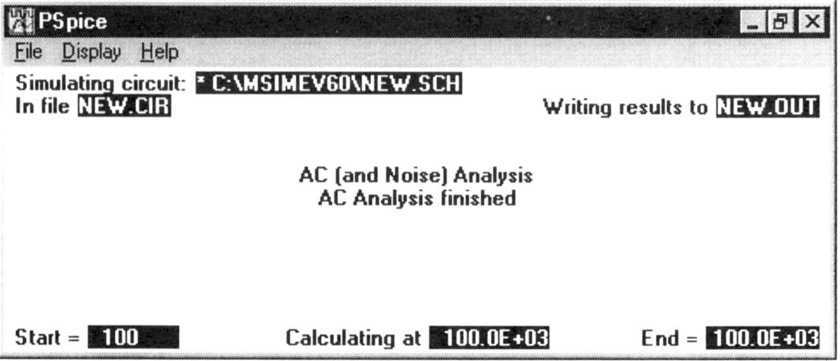

Figure F.20 Successful completion of PSpice simulation.

F.9 Plotting graphs

The Probe package will either be called automatically after a successful simulation of can be called from Psched with:

Analysis→Run Probe

When started Probe will show a graph. To add a trace the:

Trace→Add

is selected. Figure F.21 shows the Add Trace windows from the previous example. In this the trace is the voltage at Vout. Figure F.22 shows this trace.

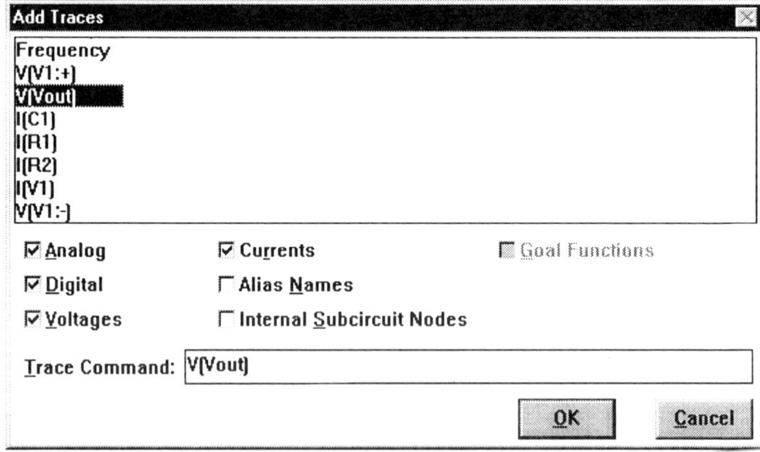

Figure F.21 Add trace window.

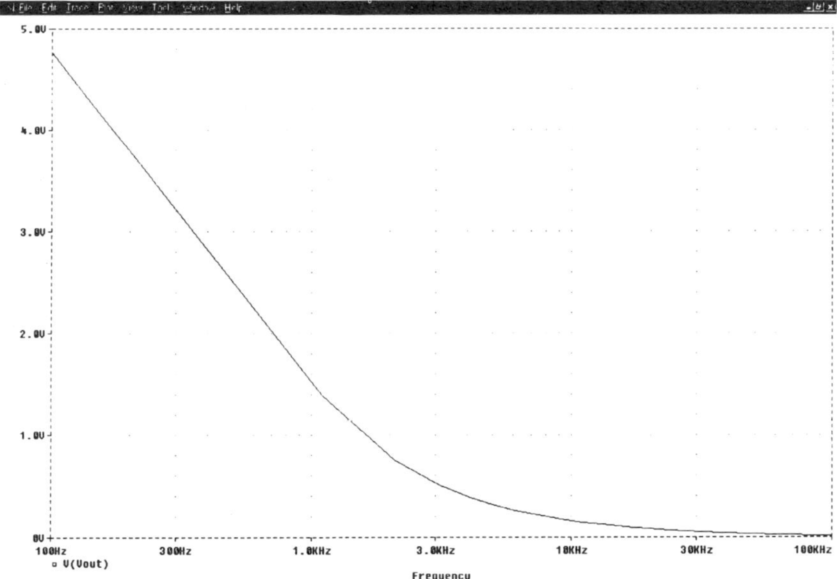

Figure F.22 Example trace.

F.10 Viewing text output

After the Probe analysis is complete the text output can be viewed. To achieve this the

Analysis→Examine Output

is selected from Psched. Listing F.4 shows a sample output of the file NEW.OUT.

Listing F.4: CONTENTS OF NEW.OUT

```
**** 09/21/96 21:18:13 *********** Evaluation PSpice (Jan 1994) *******

 * C:\MSIMEV60\NEW.SCH

****     CIRCUIT DESCRIPTION
*******************************************************************

* Schematics Version 6.0 - January 1994
* Sat Sep 21 21:18:08 1996

** Analysis setup **
.ac LIN 101 100 100k
.OP
```

```
* From [SCHEMATICS NETLIST] section of msim.ini:
.lib nom.lib

.INC "NEW.net"

**** INCLUDING NEW.net ****
* Schematics Netlist *


R_R1      $N_0001 Vout 10k
R_R2      0 Vout 10k
C_C1      0 Vout 100n
V_V1      $N_0001 0  AC 10

**** RESUMING NEW.CIR ****
.INC "NEW.als"

**** INCLUDING NEW.als ****
* Schematics Aliases *

.ALIASES
R_R1         R1(1=$N_0001 2=Vout )
R_R2         R2(1=0 2=Vout )
C_C1         C1(1=0 2=Vout )
V_V1         V1(+=$N_0001 -=0 )
_    _(Vout=Vout)
.ENDALIASES


**** RESUMING NEW.CIR ****

.probe


.END

**** 09/21/96 21:18:13 *********** Evaluation PSpice (Jan 1994) *******

 * C:\MSIMEV60\NEW.SCH


 ****     SMALL SIGNAL BIAS SOLUTION       TEMPERATURE =   27.000 DEG C
**********************************************************************

 NODE   VOLTAGE     NODE   VOLTAGE     NODE   VOLTAGE     NODE   VOLTAGE
( Vout)    0.0000  ($N_0001)    0.0000


    VOLTAGE SOURCE CURRENTS
    NAME         CURRENT

    V_V1         0.000E+00

    TOTAL POWER DISSIPATION   0.00E+00  WATTS


**** 09/21/96 21:18:13 *********** Evaluation PSpice (Jan 1994) *******
 * C:\MSIMEV60\NEW.SCH
****     OPERATING POINT INFORMATION      TEMPERATURE =   27.000 DEG C
**************************************************************************
          JOB CONCLUDED
          TOTAL JOB TIME            .56
```

F.11 Exercises

F.1 Figure F.23 shows a simple common-emitter amplifier. The results of a bias point simulation are:

Vb=1.1674 V, Ve=0.5395 V and Vout=6.6411 V

Simulate the circuit and prove these results. Show also that the current leaving V1 is 0.6563 mA and the total power dissipation is 7.88 mW.

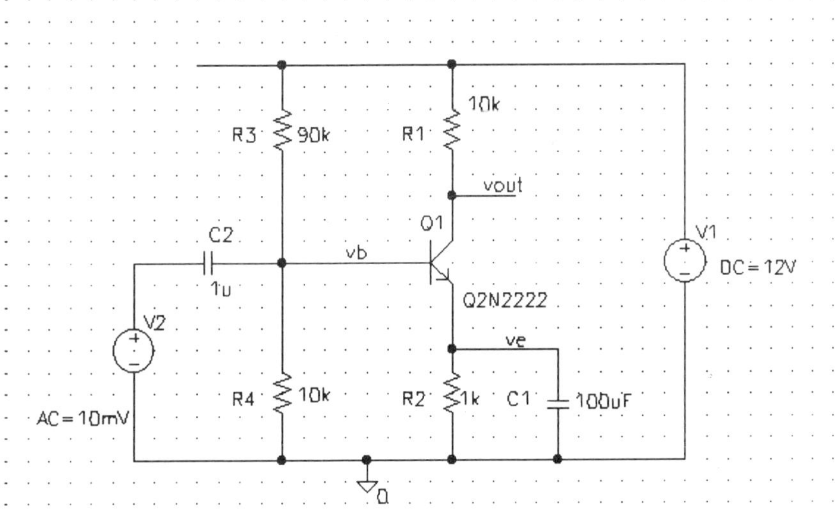

Figure F.23 Simple common-emitter amplifier.

F.2 Figure F.24 shows an AC simulation plot of the output of the circuit at Vout. Prove this result and thus determine the AC gain.

F.3 The circuit in Figure F.25 is a modified version of the common-emitter amplifier in Question F.1. It now contains a single frequency voltage source of 10 kHz which has an amplitude of 10 mV and an offset of 0 V. Simulate this circuit for its transient response and show that the response from 0 to 1 ms is given in Figure F.26. Note that the transient set up window is shown in Figure F.27.

F.4 Figure F.28 shows an active low pass filter with an LF411 op-amp. Enter the circuit in PSpice and show that the resultant plot of the output (Vout) is given in Figure F.29.

F.5 Modify the schematic in F.1 so that the supplies to the op-amp are +10V (connected to V+) and –10V (connected to V–).

F.6 Swap the resistor and capacitor in the circuit in Figure F.28 and show that it acts as a high-pass filter.

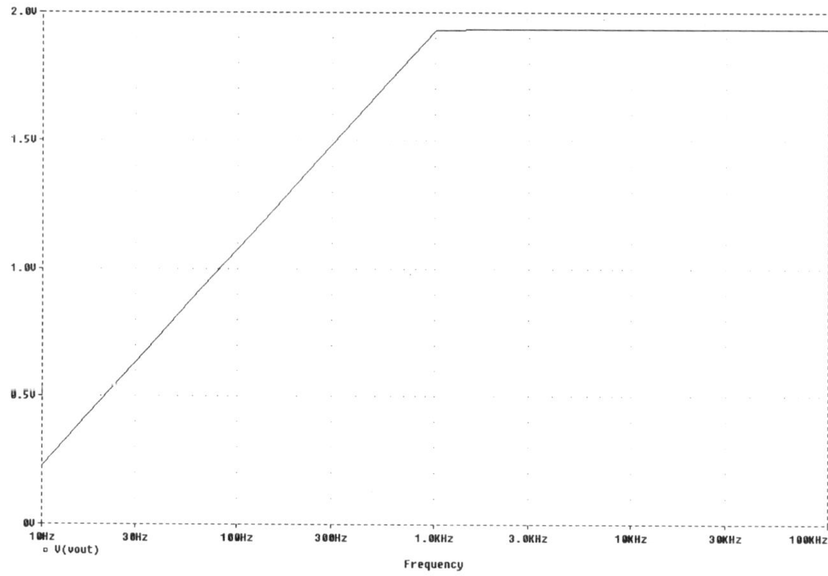

Figure F.24 AC frequency response of common-emitter amplifier.

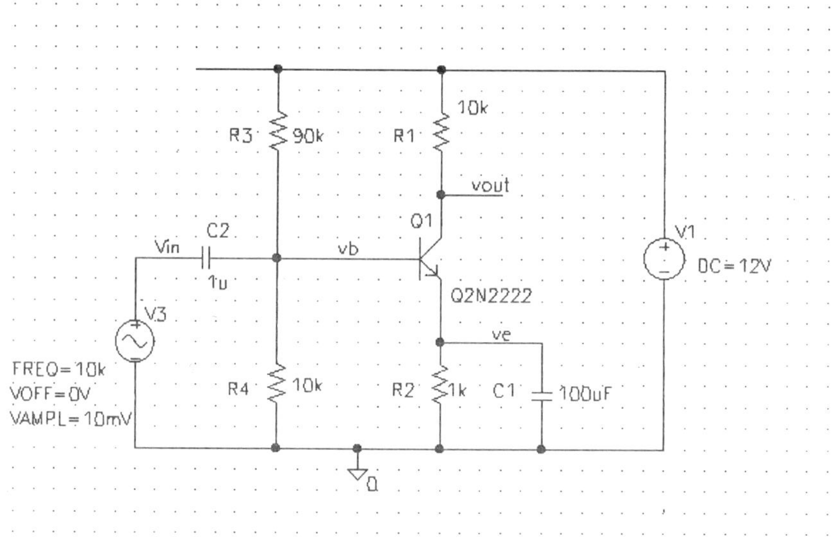

Figure F.25 Common-emitter amplifier with a single frequency source of 10 kHz.

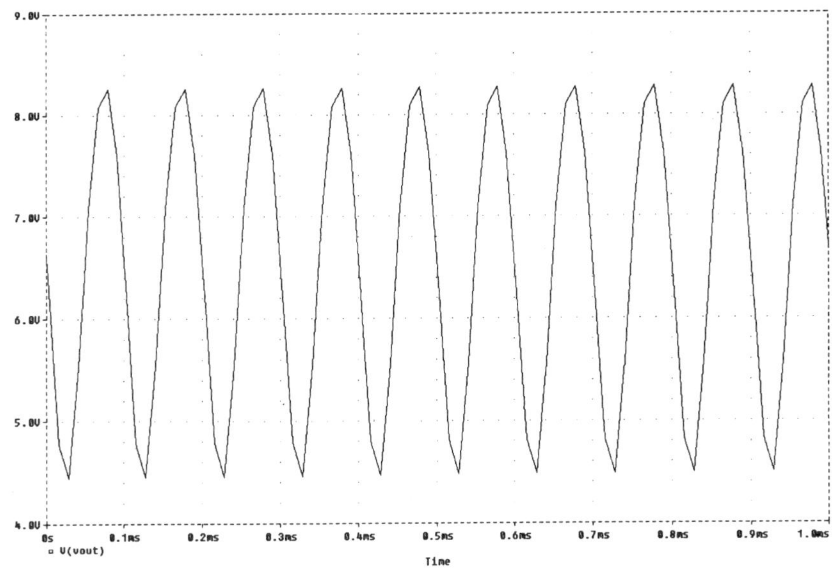

Figure F.26 Common-emitter amplifier transient response.

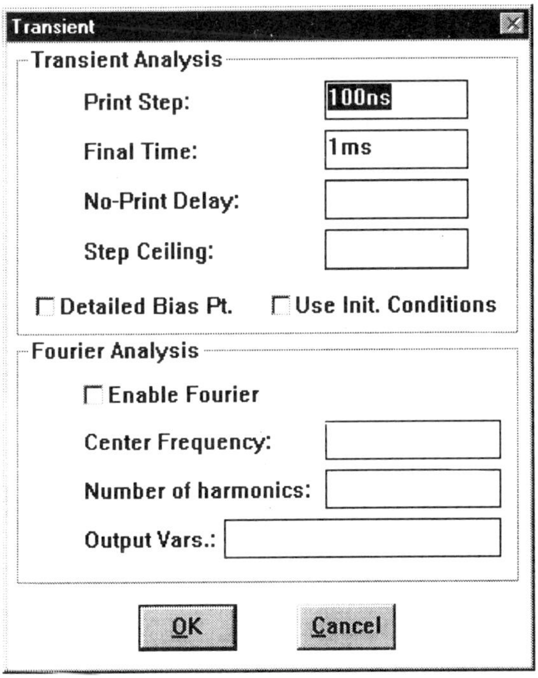

Transient

Transient Analysis

Print Step: 100ns

Final Time: 1ms

No-Print Delay:

Step Ceiling:

Detailed Bias Pt. Use Init. Conditions

Fourier Analysis

Enable Fourier

Center Frequency:

Number of harmonics:

Output Vars.:

OK Cancel

Figure F.27 Transient set up window.

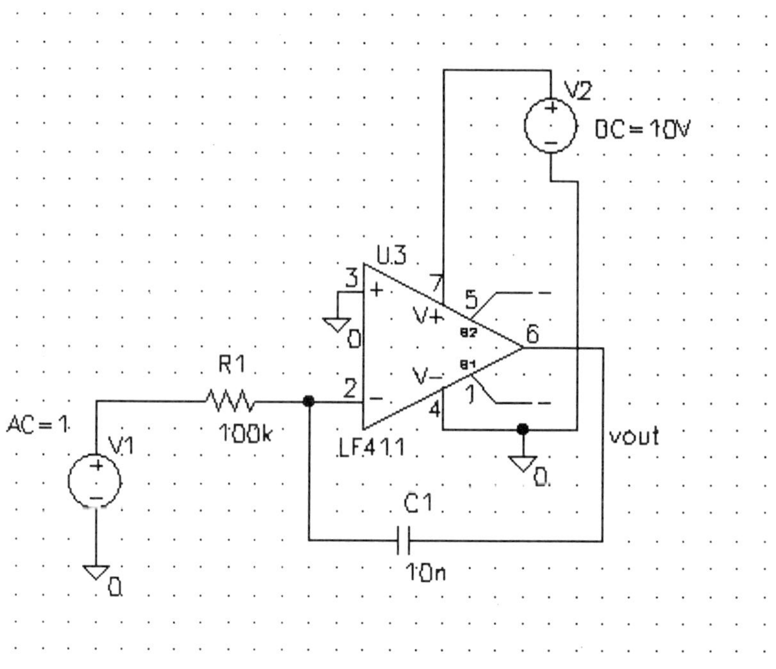

Figure F.28 Op-amp low pass filter.

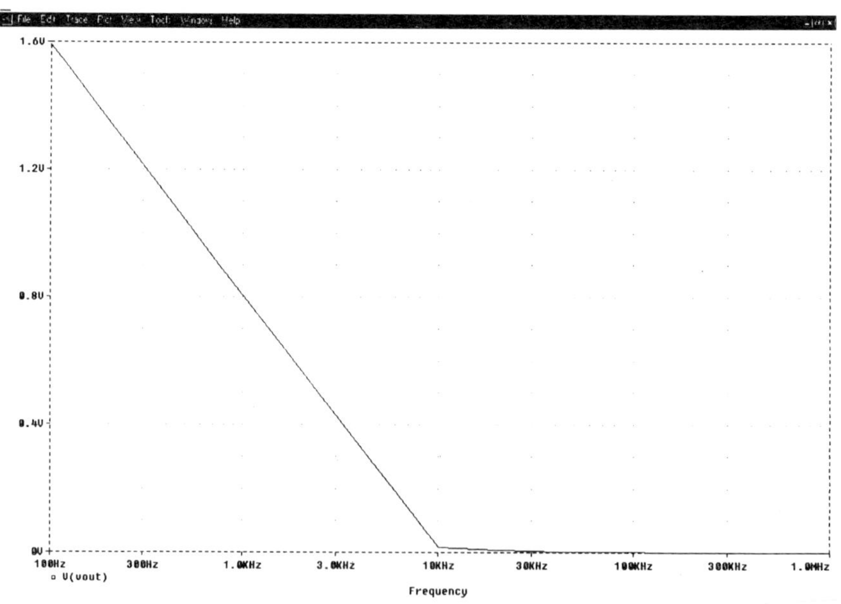

Figure F.29 Example trace.

PSpice Reference

G.1 Introduction

This appendix contains a reference section for PSpice components.

G.2 GaAsFET (B)

GaAs FETs are used in many high-frequency applications, especially in high-speed digital circuit and microwave applications. The Spice model for a GaAs FET is an intrinsic FET with an ohmic resistance (RD/area) in series with the drain, another ohmic resistance (RS/area) in series with the source, and another ohmic resistance (RG) in series with the gate. The area value is the relative device area . There are three main models used, these are:

- "Curtice" model (LEVEL=1)
- "Raytheon" or "Statz" model (Level=2);
- "TriQuint" model (level=3)).

The general form is:

```
B<name> <drain node>  <gate node>  <source node>
+ <model name>  [area value]
```

Examples

```
B1   12   13   14      GMODEL
B2   22   14   23      GMODEL2
```

Table G.1 gives the parameters used in the model. In general it is specified by:

```
.MODEL    <model name> GASFET [model parameters]
```

For example

```
.MODEL GMODEL B(level=2 vto=-1.5)
```

Table G.1 GaAs parameters.

Model parameters	*Description*	*Units*	*Default*
LEVEL	Model index (1, 2, or 3)		1
VTO	Pinch-off voltage	V	–2.5
ALPHA	Saturation voltage parameter	V^{-1}	2.0
BETA	Transconductance coefficient	A/V^{-2}	0.1
B	Doping tail extending parameter (LEVEL=2 only)	V^{-1}	0.3
LAMBDA	Channel-length modulation	V^{-1}	0
GAMMA	Static feedback parameter (LVL=3 only)		0
DELTA	Output feedback parameter (LVL=3 only)	$(A\cdot V)^{-1}$	0
Q	Power-law parameter (LEVEL=3 only)		2
TAU	Conduction current delay time	s	0
RG	Gate ohmic resistance	Ω	0
RD	Drain ohmic resistance	Ω	0
RS	Source ohmic resistance	Ω	0
IS	Gate p-n saturation current	A	10^{-14}
N	Gate p-n emission coefficient		1
M	Gate p-n grading coefficient		0.5
VBI	Gate p-n potential	V	1.0
CGD	Zero-bias gate-drain p-n capacitance	F	0
CGS	Zero-bias gate-source p-n capacitance	F	0
CDS	Drain-source capacitance	F	0
FC	Forward-bias depletion capacitance coefficient		0.5
VDELTA		V	0.2
VMAX	Capacitance limiting voltage (LVL 2 & 3)	V	0.5
EG	Bandgap voltage (barrier height)	eV	1.11
XTI	IS temperature exponent		0
VTOTC		volt/°C	0
BETATCE		%/°C	0
TRG1	RG temperature coefficient (linear)	$°C^{-1}$	0
TRD1	RD temperature coefficient (linear)	$°C^{-1}$	0
TRS1	RS temperature coefficient (linear)	$°C^{-1}$	0
KF	Flicker noise coefficient		0
AF	Flicker noise exponent		1

G.3 Capacitor (C)

Spice allows for a fixed capacitance or a variable capacitance which is controlled by the applied voltage. With a variable capacitance the *initial value* is the initial guess for the voltage across the capacitor during the bias point calculation. The general form is:

```
C<name>  <(+) node>  <(-) node>  [model name]  <value>
+        [IC=<initial value>]
```

Examples

```
C3      15   0    1uF
C10     1    2    0.2E-12   IC=2.5V
CIN     4    0    CMODEL    100pF
```

Model Form

```
.MODEL   <model name> CAP [model parameters]
```

Table G.2 Capacitor parameters.

Model parameters	*Description*	*Units*	*Default*
C	Capacitance multiplier		1
VC1	Linear voltage coefficient	V^{-1}	0
VC2	Quadratic voltage coefficient	V^{-2}	0
TC1	Linear temperature coefficient	$°C^{-1}$	0
TC2	Quadratic temperature coefficient	$°C^{-2}$	0

G.4 Diode (D)

The diode is modelled as an ohmic resistance (RS/area) in series with an intrinsic diode. An <(+) node> connects to the anode and <(–) node> to the cathode. A positive current flows from the anode through the diode to the cathode. The general form is:

```
D<name>  <(+) node>  <(-) node>  <model name>
+ [area value]
```

Examples

```
DCLAMP 14  0 DMOD
D13    15 17 SWITCH 1.5
```

Model Form

```
.MODEL   <model name> D [model parameters]
```

An example model for the D1N750 diode is:

```
.model D1N750 D(Is=880.5E-18 Rs=.25 Ikf=0 N=1 Xti=3
+ Eg=1.11 Cjo=175p M=.5516 Vj=.75 Fc=.5 Isr=1.859n
+ Nr=2 Bv=4.7 Ibv=20.245m Nbv=1.6989 Ibvl=1.9556m
+ Nbvl=14.976 Tbvl=-21.277u)
```

Table G.3 Diode parameters.

Model parameters	*Description*	*Units*	*Default*
IS	Saturation current	A	10^{-14}
N	Emission coefficient		1
ISR	Recombination current parameter	A	0
NR	Emission coefficient for ISR		2
IKF	High-injection "knee" current	A	Infinite
BV	Reverse breakdown "knee" voltage	V	Infinite
IBV	Reverse breakdown "knee" current	A	10^{-10}
NBV	Reverse breakdown ideality factor		1
IBVL	Low-level reverse breakdown "knee" current	A	0
NBVL	Low-level reverse breakdown ideality factor		1
RS	Parasitic resistance	Ω	0
TT	Transit time	s	0
CJO	Zero-bias p-n capacitance	F	0
VJ	p-n potential	V	1
M	p-n grading coefficient		0.5
FC	Forward-bias depletion capacitance coefficient		0.5
EG	Bandgap voltage (barrier height)	eV	1.11
XTI	IS temperature exponent		3
TIKF	IKF temperature coefficient (linear)	$°C^{-1}$	0
TBV1	BV temperature coefficient (linear)	$°C^{-1}$	0
TBV2	BV temperature coefficient (quadratic)	$°C^{-2}$	0
TRS1	RS temperature coefficient (linear)	$°C^{-1}$	0
TRS2	RS temperature coefficient (quadratic)	$°C^{-2}$	0
KF	Flicker noise coefficient		0
AF	Flicker noise exponent		1

G.5 Voltage-controlled voltage source (E)

The voltage controlled voltage source (VCVS) is used to simulate a voltage controlled voltage source. Controlling functions include simple gain, polynomial functions, table generated, chebyshev function, and frequency dependent. The general forms are:

```
E<name>  <(+) node>  <(-) node>
+  <(+) controlling node>  <(-) controlling node>  <gain>

E<name>  <(+) node>  <(-) node>  POLY(<value>)
+     < <(+) controlling node>  <(-) controlling node> >*
+     < <polynomial coefficient value> >*

E<name>  <(+) node>  <(-) node> VALUE = { <expression> }
```

```
E<name>  <(+) node> <(-) node> TABLE { <expression> } =
+   < <input value>,<output value> >*

E<name>  <(+) node> <(-) node> LAPLACE { <expression> } =
+   { <transform> }

E<name>  <(+) node> <(-) node> FREQ { <expression> } =
+   < <frequency value>,<magnitude value>,<phase value> >*

E<name>  <(+) node> <(-) node> CHEBYSHEV {<expression> } =
+   <type>,<cutoff frequencies>*,<attenuation>*
```

Examples

```
EIN        4   5  10 11   1.0
ESRC      10   0  POLY(1) 13   0 0 450
EVAL      20   2  POLY(2) 4 0  5  0  0.1  10.6 0.15 0.05
EFUNC      3   0  VAL = {7V*SQRT(V(3,2))}
ETAB       6   0  TABLE {V(ANODE,CATHODE)} = (0,0) (30,1)
ERC_CIR   15   0  LAPLACE {V(10)} = {1/(1+.002*s)}
EFILTER    6   0  CHEBYSHEV {V(10)} = LP 8k 10K .3dB 30dB
```

G.6 Current-controlled current source (F)

The current-controlled current source (CCCS) is similar to the VCVS but the name of the voltage source is defined by a name. The genenal forms are:

```
F<name>  <(+) node>  <(-) node>
+     <controlling V device name>  <gain>

F<name>  <(+) node>  <(-) node>  POLY(<value>)
+     <controlling V device name>
+     < <polynomial coefficient value> >
```

Examples

```
FINI 1   2 VSENSE 10.0
FAMP     13    0 POLY(1) VIN 0 500
FNONLIN 100 101 POLY(2) VCNTRL1 VCINTRL2 0.0 13.6 0.2
+ 0.005
```

The value within the braces after the POLY statement defines the number of dimensions of the polynomial.

G.7 Voltage-controlled current source (G)

The voltage-controlled current source is similar to the VCVS. The general forms are:

```
G<name>  <(+) node>  <(-) node>
+     <(+) controlling node>  <(-) controlling node>
+     <transconductance>

G<name>  <(+) node>  <(-) node>  POLY(<value>)
+   < <(+) controlling node>  <(-) controlling node> >*
+     <polynomial coefficient value>*

G<name>  <(+) node> <(-) node> VALUE = { <expression> }

G<name>  <(+) node> <(-) node> TABLE { <expression> } =
+     < <input value>,<(output) value> >*

G<name> <(+) node> <(-) node> LAPLACE { <expression> } =
+     { <transform> }

G<name>  <(+) node> <(-) node> FREQ { <expression> } =
+ < <frequency value>,<magnitude value>,<phase value> >*

G<name> <(+) node> <(-) node> CHEBYSHEV {<expression>} =
+     <type>,<cutoff frequencies>*,<attenuation>*
```

Examples

```
GIN    3    3  11 12  2.0
GAMP   10   0  POLY(1)   22  0 0 500
GLIN   10   11 POLY(2) 3 0 4 0 0.0 13.6 0.2 0.005
GPSK   11   6  VALUE = {5MA*SIN(6.28*10kHz*TIME+V(3))}
GTT    ANODE CATHODE VALUE = {200E-6*PWR(V(1)*V(2),1.5)}
GLOSS  5 0   LAPLACE {V(10)} = {exp(-sqrt(C*s*(R+L*s)))}
```

G.8 Current-controlled voltage source (H)

The current-controlled voltage source is similar to the CCCS. The genenal forms are:

```
H<name>  <(+) node>  <(-) node>
+     <controlling V device name>  <transresistance>

H<name>  <(+) node>  <(-) node>  POLY(<value>)
+     <controlling V device name>*
+     <polynomial coefficient value>*
```

Examples

```
HSENSE 1  2 VSENSE 10.0
HSRC   13 0 POLY(1) VIN 0 500
HIN    1  2 POLY(2) VCNTRL1 VCINTRL2 0.0 13.6 0.2 0.005
```

G.9 Independent current source and stimulus (I)

The independent current source has a positive current flowing from the (+) node through the source to the (–) node. Types of source include exponential, pulse, piecewise linear, frequency modulated and sinusoidal. IBIAS drives node 13 to have a negative voltage. The default value is zero for the DC, AC, and transient values. None, any, or all of the DC, AC, and transient values may be specified. The AC phase value is in degrees.General Form

```
I<name>  <(+) node>  <(-) node>
+        [ [DC] <value> ]
+        [ AC <magnitude value> [phase value] ]
+        [transient specification]
```

Examples

```
IBIAS  13  0 1.5mA
IAC    2   3 AC .01
IACPHS 2   3 AC .01 45
IPULSE 1   0 PULSE(-1mA 1mA 2ns 2ns 2ns 50ns 100ns)
I3     2   7 DC .002  AC 1  SIN(.002 .002 1.5MEG)
```

If present, the [transient specification] must be one of:

EXP (<parameters>) for an exponential waveform
PULSE (<parameters>) for a pulse waveform
PWL (<parameters>) for a piecewise linear waveform
SFFM (<parameters>) for a frequency-modulated waveform
SIN (<parameters>) for a sinusoidal waveform

Refer to section 12.6 for the format of these transient types.

G.10 Junction FET (J)

The JFET is modelled as an intrinsic FET with an ohmic resistance (RD/area) in series with the drain, and with another ohmic resistance (RS/area) in series with the source. Positive current is current flowing into a terminal. The area value is the relative device area and defaults to 1. The general form is:

```
J<name>  <drain node>  <gate node>  <source node>
+        <model name>  [area value]
```

Examples

```
JIN 12 12  2  JMODEL
J1  1  2   3  J1   2.0
```

Model forms

```
.MODEL   <model name> NJF [model parameters]
.MODEL   <model name> PJF [model parameters]
```

The model for the J2N3819 JFET is:

```
.model J2N3819 NJF(Beta=1.304m Betatce=-.5 Rd=1
+ Rs=1 Lambda=2.25m Vto=-3 Vtotc=-2.5m Is=33.57f
+ Isr=322.4f N=1 Nr=2 Xti=3 Alpha=311.7
+  Vk=243.6 Cgd=1.6p M=.3622 Pb=1 Fc=.5 Cgs=2.414p
+ Kf=9.882E-18 Af=1)
```

Table G.4 JFET parameters.

Model parameters	*Description*	*Units*	*Default*
VTO	Threshold voltage	V	–2.0
BETA	Transconductance coefficient	A/V^2	10^{-4}
LAMBDA	Channel-length modulation	V^{-1}	0
IS	Gate p-n saturation current	A	10^{-14}
N	Gate p-n emission coefficient		1
ISR	Gate p-n recombination current parameter	A	0
NR	Emission coefficient for ISR		2
ALPHA	Ionization coefficient	V^{-1}	0
VK	Ionization 'knee' voltage	V	0
RD	Flicker noise coefficient		0
AF	Flicker noise exponent		1

G.11 Inductor-coupling (transformer core) (K)

Inductor coupling (transformer core) couples two, or more, inductors. The general form is:

```
K<name>  L<inductor name>  < L<inductor name> >*
+        <coupling value>
K<name>  < L<inductor name> >*  <coupling value>
+        <model name>  [size value]
```

The <coupling value> is the 'coefficient of mutual coupling' and must be between 0 and 1. This coefficient is defined by the equation
Examples

```
KCKT      L1  L2 0.8
KTRANS    LPRIM  LSEC 0.95
K1        L1  L2  L3  L4 0.92 KMODEL
```

Model form

```
.MODEL   <model name> CORE [model parameters]
```

Table G.5 Inductor-coupling parameters.

Model parameters	*Description*	*Units*	*Default*
LEVEL	Model index		1
AREA	Mean magnetic cross-section	cm^2	0.1
PATH	Mean magnetic path length	cm	1.0
GAP	Effective air-gap length	cm	0
PACK	Pack (stacking) factor		1.0
MS	Magnetization saturation	amp/meter	10^6
A	Thermal energy parameter	amp/meter	10^3
C	Domain flexing parameter		0.2
K	Domain anisotropy parameter	amp/meter	500
ALPHA	Interdomain coupling parameter(LEVEL=1)	10^{-3}	
GAMMA	Domain damping parameter (LEVEL=1)	sec^{-1}	infinite

G.12 Inductor (L)

The inductor is connected between the (+) and (–) nodes. A positive current flows from the (+) node through the inductor to the (–) node. The general form is:

```
L<name>  <(+) node>  <(-) node>  [model name]  <value>
+        [IC=<initial value>]
```

Examples

```
LLOAD  15   0 10mH
L2      1   2 1E-3
LCHOKE  1  0   LMODEL 0.02
LSENSE  4  3   6uH     IC=10mA
```

Model Form

```
.MODEL   <model name> IND [model parameters]
```

Table G.6 Inductor parameters.

Model parameters	*Description*	*Units*	*Default*
L	Inductance multiplier	1	
IL1	Linear current coefficient	A^{-1}	0
IL2	Quadratic current coefficient	A^{-1}	0
TC1	Linear temperature coefficient	$°C^{-1}$	0
TC2	Quadratic temperature coefficient	$°C^{-2}$	0

G.13 MOSFET (M)

MOSFETs are modelled as an intrinsic MOSFET with ohmic resistances in series with the drain, source, gate, and bulk (substrate). There is also a shunt resistance (RDS) in parallel with the drain-source channel.

PSpice provides four MOSFET device models, which differ in the formulation of the I-V characteristic, these are:

- the Shichman-Hodges model (LEVEL=1);
- geometry-based, analytic model (LEVEL=2);
- semi-empirical, short-channel model (LEVEL=3);
- the BSIM model (LEVEL=4).

The general form is:

```
M<name>  <drain node>  <gate node>  <source node>
+     <bulk/substrate node>  <model name>
+     [L=<value>]  [W=<value>] [AD=<value>]  [AS=<value
+     [PD=<value>] [PS=<value>] [NRD=<value>]
+     [NRS=<value>] [NRG=<value>]  [NRB=<value>]
+     [M=<value>]
```

Where L and W are the channel length and width. AD and AS are the drain and source diffusion areas, and PD and PS are the drain and source diffusion perimeters. The drain-bulk and source-bulk saturation currents are specified either by JS, which is multiplied by AD and AS, or by IS, which is an absolute value. The zero-bias depletion capacitances are specified by CJ, which is multiplied by AD and AS, and by CJSW, which is multiplied by PD and PS. Or they can be set by CBD and CBS, which are absolute values.

NRD, NRS, NRG, and NRB are the relative resistivities of the drain, source, gate, and substrate in squares. These parasitic (ohmic) resistances can be specified either by RSH, which is multiplied by NRD, NRS, NRG, and NRB, respectively, or by RD, RS, RG, and RB, which are absolute values.

PD and PS default to 0, NRD and NRS default to 1, and NRG and NRB default to 0. Defaults for L, W, AD, and AS may be set in the .OPTIONS statement. If AD or AS defaults are not set, they also default to 0. If L or W defaults are not set, they default to 100u.

The device multiplier, M, simulates the effect of multiple devices in parallel and has a default value of 1. Effective widths, overlap and junction capacitances, and junction currents of the MOSFET are multiplied by M and the parasitic resistance values (such as RD and RS) are divided by M.

Examples

```
M1   1 2 3    0 PMODEL  L=25u W=12u
M3   1 3 0    0 PMOD
M1   2 3 0    0 PMOD2 M=2
M12  0 2 10   100 NMOD L=33u W=12u AD=288p AS=288p
+ PD=60u PS=60u NRD=14 NRS=24 NRG=10
```

Model forms

```
.MODEL   <model name> NMOS [model parameters]
.MODEL   <model name> PMOS [model parameters]
```

An example model for the IRF150 device is:

```
.model IRF150 NMOS(Level=3 Gamma=0 Delta=0 Eta=0 Theta=0
+ Kappa=0 Vmax=0 Xj=0 Tox=100n Uo=600 Phi=.6 Rs=1.624m
+ Kp=20.53u W=.3 L=2u Vto=2.831 Rd=1.031m Rds=444.4K
+ Cbd=3.229n Pb=.8 Mj=.5 Fc=.5 Cgso=9.027n Cgdo=1.679n
+ Rg=13.89 Is=194E-18 N=1 Tt=288n)
```

G.14 Digital input (N)

Refer to PSpice manual.

G.15 Digital output (O)

Refer to PSpice manual.

G.16 Bi-polar transistor (Q)

The bi-polar transistor is modelled as an intrinsic transistor with resistances in series with the collector (RC/area), the base (value varies with current), and with the emitter (RE/area). A positive current flow is into a terminal. The area value is the relative device area and defaults to 1.

A substrate node is optional, and if not specified it defaults to ground. For model types NPN and PNP, the isolation junction capacitance is connected between the intrinsic-collector and substrate nodes. The general form is:

```
Q<name>  <collector node>  <base node>  <emitter node>
+      [substrate node]  <model name>  [area value]
```

Examples

```
Q1   14 2 13 PNPNOM
Q13 15 3  0  1 NPNSTRONG 1.5
Q7   VC 5 12 [SUB] LATPNP
```

Model forms

```
.MODEL   <model name> NPN [model parameters]
.MODEL   <model name> PNP [model parameters]
.MODEL   <model name> LPNP [model parameters]
```

Table G.7 Bi-polar transistor parameters.

Model parameters	*Description*	*Units*	*Default*
IS	Transport saturation current	A	10^{-16}
BF	Ideal maximum forward beta		100
NF	Forward current emission coefficient		1
VAF	Forward Early voltage	V	infinite
IKF	Corner for forward-beta high-current roll-off	A	infinite
ISE	Base-emitter leakage saturation current	A	0
NE	Base-emitter leakage emission coefficient		1.5
BR	Ideal maximum reverse beta		1
NR	Reverse current emission coefficient		1
VAR	Reverse Early voltage	V	infinite
IKR	Corner for reverse-beta high-current roll-off	A	infinite
ISC	Base-collector leakage saturation current	A	0
NC	Base-collector leakage emission coefficient		2
NK	High-current roll-off coefficient		.5
ISS	Substrate p-n saturation current	A	0
NS	Substrate p-n emission coefficient		1
RE	Emitter ohmic resistance	Ω	0
RB	Zero-bias (maximum) base resistance	Ω	0
RBM	Minimum base resistance	Ω	RB
IRB	Current at which Rb falls halfway to RBM	Ω	infinite
RC	Collector ohmic resistance	Ω	0
CJE	Base-emitter zero-bias p-n capacitance	F	0
VJE	Base-emitter built-in potential	V	0.75
MJE	Base-emitter p-n grading factor		0.33
CJC	Base-collector zero-bias p-n capacitance	F	0
VJC	Base-collector built-in potential	V	0.75
MJC	Base-collector p-n grading factor		0.33
XCJC	Fraction of Cbc connected internal to Rb		1
CJS	Substrate zero-bias p-n capacitance	F	0
VJS	Substrate p-n built-in potential	V	0.75
MJS	Substrate p-n grading factor		0

Table G.7 (Cont.).

Model parameters	*Description*	*Units*	*Default*
FC	Forward-bias depletion capacitor coefficient		0.5
TF	Ideal forward transit time	s	0
XTF	Transit time bias dependence coefficient		0
VTF	Transit time dependency on Vbc	V	infinite
ITF	Transit time dependency on Ic	A	0
PTF	Excess phase at $1/(2\pi \times \text{TF})$Hz	°	0
TR	Ideal reverse transit time	s	0
QCO	Epitaxial region charge factor	C	0
RCO	Epitaxial region resistance	Ω	0
VO	Carrier mobility "knee" voltage	V	10
GAMMA	Epitaxial region doping factor		10^{-11}
EG	Bandgap voltage (barrier height)	eV	1.11
XTB	Forward and reverse beta temperature coefficient		0
XTI	IS temperature effect exponent		3
TRE1	RE temperature coefficient (linear)	$°C^{-1}$	0
TRE2	RE temperature coefficient (quadratic)	$°C^{-2}$	0
TRB1	RB temperature coefficient (linear)	$°C^{-1}$	0
TRB2	RB temperature coefficient (quadratic)	$°C^{-2}$	0
TRM1	RBM temperature coefficient (linear)	°C^-1	0
TRM2	RBM temperature coefficient (quadratic)	°C^-2	0
TRC1	RC temperature coefficient (linear)	$°C^{-1}$	0
TRC2	RC temperature coefficient (quadratic)	$°C^{-1}$	0
KF	Flicker noise coefficient		0
AF	Flicker noise exponent		1

An example model for the Q2N2222 device is:

```
.model Q2N2222  NPN(Is=14.34f Xti=3 Eg=1.11 Vaf=74.03
+ Bf=255.9 Ne=1.307 Ise=14.34f Ikf=.2847 Xtb=1.5 Br=6.092
+ Nc=2 Isc=0 Ikr=0 Rc=1 Cjc=7.306p Mjc=.3416 Vjc=.75 Fc=.5
+ Cje=22.01p Mje=.377 Vje=.75 Tr=46.91n Tf=411.1p Itf=.6
+ Vtf=1.7 Xtf=3 Rb=10)
```

G.17 Resistor (R)

The (+) and (–) nodes define the polarity when the resistor has a positive voltage across it. Positive current flows from the (+) node through the resistor to the (–) node. Temperature coefficients for the resistor can be specified in-line, as in the second example. If the resistor has a model specified, then the coefficients from the model are used for the temperature updates, otherwise the in-line values are used. In both cases the temperature coefficients default to zero. The general form is:

```
R<name>  <(+) node>  <(-) node>  [model name]  <value>
+        [TC = <TC1> [,<TC2>]]
```

Examples

```
RLOAD 15 0 2K
R2     1 2 2.4E4 TC=.015,-.003
RFDBCK 3 33 RMOD 10K
```

Model form

```
.MODEL   <model name> RES [model parameters]
```

The model parameter are given next.

Table G.8 Resistor parameters.

Model parameters	*Description*	*Units*	*Default*
R	Resistance multiplier		1
TC1	Linear temperature coefficient	$°C^{-1}$	0
TC2	Quadratic temperature coefficient	$°C^{-2}$	0
TCE	Exponential temperature coefficient	%/°C	0

G.18 Voltage-controlled switch (S)

The voltage-controlled switch (VCS) is a special kind of voltage-controlled resistor. Resistance values between the <(+) switch node> and <(-) switch node> depend on the voltage between the <(+) controlling node> and <(-) controlling node>. These values vary continuously between RON and ROFF. The general form is:

```
S<name>  <(+) switch node>  <(-) switch node>
+     <(+) controlling node>  <(-) controlling node>
+     <model name>
```

Examples

```
S12  3 7 2 0 SMODEL
SW1  2 3 4 9 INRELAY
```

Model form

```
.MODEL   <model name> VSWITCH [model parameters]
```

Table G.9 VCS parameters.

Model parameters	*Description*	*Units*	*Default*
RON	'On' resistance	Ω	1.0
ROFF	'Off' resistance	Ω	10^6
VON	Control voltage for 'on' state	V	1.0
VOFF	Control voltage for 'off' state	V	0.0

G.19 Transmission line (T)

The transmission line device models a transmission line with an two ports, A and B. The (+) and (–) nodes define the polarity of a positive voltage at a port. For the ideal case, Z0 is the characteristic impedance. The length of the transmission line's length is specified either by TD, a delay in seconds, or by F and NL, a frequency and a relative wavelength at F. NL defaults to 0.25 (F is then the quarter-wave frequency). Although TD and F are both shown as optional, one of the two must be specified. Examples T1, T2, and T3 all specify the same transmission ideal line.

For the case of a lossy line, LEN is the electrical length. R, L, G, and C are the per unit length values of resistance, inductance, conductance, and capacitance, respectively.The general form for an ideal line is:

```
T<name>  <A port (+) node>  <A port (-) node>
+      <B port (+) node>  <B port (-) node>
+      [model name]
+      Z0=<value>  [TD=<value>]  [F=<value> [NL=<value
```

For a lossy line the general form is:

```
T<name>  <A port (+) node>  <A port (-) node>
+      <B port (+) node>  <B port (-) node>
+      [ <model name> [electrical length value] ]
+      LEN=<value> R=<value> L=<value>
+      G=<value> C=<value>
```

Examples

```
T1   1 2 3 4  Z0=220 TD=115ns
T2   1 2 3 4  Z0=220 F=2.25MHz
T3   1 2 3 4  Z0=220 F=4.5MHz NL=0.5
T4   1 2 3 4  LEN=1 R=.311 L=.378u G=6.27u C=67.3p
T5   1 2 3 4  TMOD 1
```

Model form

```
.MODEL <model name> TRN [model parameters]
```

G.20 Digital stimuli and primitives (U)

Refer to PSpice manual.

G.21 Independent voltage source and stimulus (V)

This independent voltage source can be used to as a voltage source for DC, AC, and voltage transients. the general form is:

```
V<name>  <(+) node>  <(-) node>
+     [ [DC] <value> ]
+     [ AC <magnitude value>  [phase value] ]
+     [transient specification]
```

Examples

```
VBIAS   13  0   2.3mV
VAC      2  3   AC .001
VACPHS   2  3   AC .001 90
VPULSE   1  0   PULSE(-1mV 1mV 2ns 2ns 2ns 50ns 100ns)
V3      26 77   DC .002  AC 1  SIN(.002 .002 1.5MEG)
```

If present, the [transient specification] must be one of:

EXP (<parameters>) for an exponential waveform
PULSE (<parameters>) for a pulse waveform
PWL (<parameters>) for a piecewise linear waveform
SFFM (<parameters>) for a frequency-modulated waveform
SIN (<parameters>) for a sinusoidal waveform

Refer to section 12.6 for the format of these transient types.

G.22 Current-controlled switch (W)

The current-controlled switch (CCS) is a special kind of voltage-controlled resistor. Resistance values between the <(+) switch node> and <(–) switch node> depend on the current through <controlling V device name>. These values vary continuously between RON and ROFF. The general form is:

```
W<name>   <(+) switch node>  <(-) switch node>
+     <controlling V device name>  <model name>
```

Examples

```
W12     13 17 VC WMOD
WRESET   5  0 VRESET RELAY
```

Model Form

```
.MODEL    <model name> ISWITCH [model parameters]
```

Table G.10 CCS parameters.

Model parameters	*Description*	*Units*	*Default*
RON	'On' resistance	Ω	1.0
ROFF	'Off' resistance	Ω	10^6
ION	Control current for 'on' state	A	10^{-3}
IOFF	Control current for 'off' state	A	0.0

G.23 Sub-circuit Call (X)

Refer to PSpice reference manual.

Further Reference

H.1 WWW reference

Further information on related subjects, such as FPGAs, PSpice, GaAs design, NMOS and CMOS design, as well as up-to-date software is available on the WWW page:

`http://www.eece.napier.ac.uk/~bill_b/mbook.html`

H.2 30-pin SIMM

The 30-pin SIMM module has 10 addressing lines (A0-A9) and 8 data lines (D0-D7). The actual address is selected using the Row Address Select (RAS) and the Column Address Select (CAS). Table H.1 defines its pins.

Table H.1 30-pin SIMM connections.

Pin	*Signal name*	*Pin*	*Signal name*
1	+5V	16	D4
2	CAS	17	A8
3	D0	18	A9
4	A0	19	
5	A1	20	D5
6	D1	21	R/W
7	A2	22	GND
8	A3	23	D6
9	GND	24	
10	D2	25	D7
11	A4	26	
12	A5	27	RAS
13	D3	28	
14	A6	29	
15	A7	30	+5V

H.3 72-pin SIMM

The 72-pin SIMM module has 10 addressing lines (A0-A10) and 32 data lines (D0-D31). The actual address using four Row Address Strobe lines (RAS0-

RAS3) and four Column Address Strobe lines (CAS0-CAS3). Table H.2 defines its pins. A 72-pin SIMM also has an automatic sensing mechanism using pins 67-70. Table H.3 lists some typical settings.

Table H.2 72-pin SIMM connections.

Pin	*Signal name*	*Pin*	*Signal name*	*Pin*	*Signal name*
1	GND	25	BLOCK SEL 0	49	D8
2	D0	26	+5V	50	D24
3	D16	27	A8	51	D9
4	D1	28	A9	52	D25
5	D17	29	RAS3	53	D10
6	D2	30	RAS2	54	D26
7	A18	31		55	D11
8	+5V	32	D6	56	D27
9		33		57	D12
10	A0	34		58	D28
11	A1	35		59	+5V
12	A2	36		60	D29
13	A3	37		61	D13
14	A4	38		62	D30
15	A5	39	GND	63	D14
16	A6	40	CAS0	64	D31
17		41	CAS2	65	D15
18	D4	42	CAS3	66	BLOCK SEL 2
19	D20	43	CAS1	67	PRES DET 0
20	D5	44	RAS0	68	PRES DET 1
21	D21	45	RAS1	69	PRES DET 2
22	D6	46	BLOCK SEL 1	70	PRES DET 3
23	D22	47	R/W	71	BLOCK SEL 3
24	D7	48		72	GND

Table H.3 Presence detect pins.

PRES DET 3	*PRES DET 2*	*PRES DET 1*	*PRES DET 0*	*Signal name*
NC	NC	GND	NC	2 MB 120 ns
NC	NC	GND	GND	2 MB 70 ns
NC	GND	NC	NC	8 MB 70 ns
NC	GND	GND	NC	2 MB 80 ns
NC	GND	GND	GND	8 MB 80 ns
GND	NC	GND	NC	2 MB 85 ns
GND	NC	GND	GND	4 MB 70 ns
GND	GND	NC	NC	4 MB 80 ns
GND	GND	NC	GND	2 MB 100 ns
GND	GND	GND	NC	4 MB 80 ns
GND	GND	GND	GND	2 MB 85 ns

where NC represents not connected.

Index

&, *see* Ampersand
2's complement, 146
2D addressing, 104, 105
4000, 40, 250
6502, 18
80486, 8, 113

—A—

Actel, 118–120, 122
ADCs, 14
Address bus, 101, 102
Addressable memory, 102
Algorithms, 5, 160
ALUs, *see* Arithmetic logic units
Ampersand, 142
AND, 10, 13, 16, 30, 75, 92, 96, 116, 123, 132, 133, 138–143, 164, 203, 204, 211
ANSI, 139, 142
Apple, 18
Application specific integrated circuit, *see* ASIC
Architecture body, 161
Arithmetic logic units, 5
ASIC, 14, 15, 17, 83, 88–91, 98, 99, 115, 117
ATE, *see* Automatic test equipment
Automatic test circuits, 220
Automatic test equipment, 7, 15, 202

—B—

Beta factor, 241
BFL, 9
Bias, 21, 26, 30–33, 48, 188, 277, 282, 284, 290, 292, 293
BILBO, 226–228
Binary
  code, 154
BIOS, 103
Bi–polar transistor, 291
Bit fields, 146–148
Bitwise operator, 142
Boolean,
  general, 10, 42, 50, 96, 116, 120, 137, 138, 140–145, 151, 153, 155, 169, 170, 174, 175, 176, 204–207, 211, 212
  differences, 205, 211
  logic, 153
  sequences, 143
boolean.h, 141, 142, 153
Braces, 143
Buffering, 2
Bug fixes, 12
Buses, 5
Busicon, 17

—C—

Cache, 112
CAD, *see* Computer aided design
Cadence, 87
Capacitance, 47, 53, 54, 65, 109, 110, 188, 238–241, 249, 250, 282, 284, 291, 292, 295
Capacitor, 106, 183, 189, 238, 266, 268, 278, 282, 283, 293
Case statement, 174
CBS, 17, 290
CCCS, 285, 286
Channel length, 65, 68, 290
Channel resistance, 53–55, 73, 74, 241
Checkerboard, 231
CIF format, 251
CLB, see Configurable logic blocks

CMOS,
general, 8–10, 21, 30, 39–42, 44, 47, 49, 50, 63, 77–82, 106, 124, 125, 129, 209, 212, 245, 249, 250, 252
inverter, 77–79, 82
NAND, 79–81
NOR, 81, 212
Code
source, 141
Combinational logic, 118, 119, 138, 217
Comparison,
NAND and NOR gates, 75
Compiler, 146
Configurable logic blocks, 124–126, 128
Computer aided design, 15
Concurrent, 173, 179, 261
Configuration declaration, 161
Connections:
between cells, 91
Contact cuts, 64, 68, 245, 247, 252
Controllability, 202
Current source, 285, 287
Current sources, 187
Customization, 12, 15

—D—

DACs, 14
DC simulation, 196
DC transfer characteristics, 199
DCL, *see* Direct coupled logic
Decoders, 5, 229
Department of Defense, 159
Depletion mode, 45, 51, 63, 64, 72, 239
Design abstraction, 5, 6, 159, 160, 161
Die, 13
Dielectric constant, 237, 241
Differential amplifier, 193–195
Diffuse, 244, 246, 247
Diffusion, 63–65, 67–70, 73, 77–79, 82, 92–95, 109, 110, 122, 191, 244, 245, 249, 250, 290
DIGCHIP, 83, 88
Digital IC technologies, 8
DIL, 124
Diode, 1, 9, 21–23, 27, 28, 34, 48, 129, 183, 200, 283, 284
Direct coupled logic, 9
DON'T WALK, 99
Doping, 51, 63, 293
DOS, 18
Dot commands, 187
Drain, 39, 40, 45, 46, 51–56, 60, 63–65, 73–75, 77, 123, 184, 191, 192, 237–241, 281, 282, 287, 290
DRAM, 101, 106–113, 234
DRC, 15
Dynamic RAM, *see* DRAM

—E—

ECAD, 7, 10, 87, 201, 207
ECL, 8, 10, 31, 32, 34–38
EEPROM, 103, 121–123, 133
EEPROMs, 3, 11, 116
Emiter coupled logic, *see* ECL
Electrical characteristics, 2, 29, 171
Electronic computer aided design, *see* ECAD
Electrons, 1, 123
Enhanced mode, 51, 69, 237
Entity declaration, 161
EPROM, 103, 116, 121–123, 129–131, 133
EPROMs, 11, 103, 116, 130
Etch, 244–247
Exponential
transient, 185
waveform, 185, 287, 296
Exclusive–OR, 203

—F—

Fabrication, 243, 245, 251
FAST, 9
Fault,
coverage, 202
finding, 5, 201
modelling, 201, 202
Field programmable, 11, 12, 15, 116
Field Programmable Gate Arrays, *see* FPGAs
File,
text, 151
Firmware, 3, 4
Fixed array, 116
Fleming, 1
Forward biased, 24, 28

Foundry, 15
FPGAs, 115, 116, 120, 122, 124, 132, 133
FPLAs, 11
Full-custom, 7, 10, 15, 115
Function macros, 141
Fusible links, 11, 116

—G—

GaAs, 7, 9, 10, 184, 281, 282
Gallium Arsenide, *see* GaAs
Gate,
  general, 22, 24–81, 92–98, 106, 110, 115, 116, 119, 121, 123, 133, 153, 154, 161, 171, 172, 178, 184, 191, 192, 202–206, 209, 210, 211, 212, 219, 222, 227, 237–241, 242, 281, 282, 290
  arrays, 11, 12, 15, 116
  capacitance, 53, 110, 239, 240, 241
  logic, 3, 11–14, 24, 26, 27, 29, 30, 35-40, 43–49, 53, 58–60, 63, 73, 75, 82, 92, 96, 116, 117, 120, 122, 124, 129, 132, 133, 153, 191, 202, 205, 211, 219, 222
  transfer, 32
Geometric shapes, 251
Germanium, 1, 17
Gray code, 154
GREEN, 64, 82, 83, 86, 87, 99, 148–151, 157
GSI, 3

—H—

Hard disk, 101, 103
Hardware, 3, 4, 7, 12, 18, 87, 159, 161, 171
HDL, 16
Header file, 141
HMOS, 9, 10
Hughes, 17
Hybrid IC, 16

—I—

I^2L, 8, 10, 29, 30, 36, 37
IBM, 17, 18, 219
IC
  general, 1, 14–16, 17, 83, 115, 159, 251
  design, 3, 7, 15
  fabrication, 243
IEEE 1076 standard, 159
If statement, 174
if(), 140
IIL, *see* I^2L
Implementation, 88
impurity, 51, 245
Inductor, 288–290
Initialization, 143
Input/output blocks, *see* IOBs
Input/output requirements, 83
Insulator, 60, 240, 245
Integrated circuit, *see* IC
Intel, 17, 18, 63, 113
Interference, 2, 234
Inverter,
  general, 69, 70, 165
  ratio, 57–60
IOBs, 125, 128
Ion implant, 63, 64
Isolation, 2, 7, 51, 291

—J—

Jargon, 15
JFET, 287, 288

—L—

L:W, 58, 59, 69, 71, 72, 82
Lambda,
  general, 66, 67, 70, 73, 249
  design rules, 66, 67, 70, 73
Layers, 251
LCC, 125
Level–sensitive scan design, 219, 220
LFSR, 222, 223, 225–227
LOCST, *see* LSSD on-chip self-test
Logic cell design, 92
Logic simulation, 144
Logical operator, 138
Loop statement, 177
Lorentz, 1
LS, 9, 36

LSI, 1, 3
LSSD, *see* Level-sensitive scan design
LSSD on-chip self-test, 19

—M—

Marching 0's, 230, 236
Marching 1's, 230, 236
Mask, 5, 64, 65, 94, 95, 148, 160, 243–247
Mask-programmable, 115, 116
Mead and Conway, 66
Memory,
  addressing size, 102
  decoders failures, 229
  multiple writes, 229
  pattern sensitivity, 229
  slow sense amplifier, 229
  systems, 101
  test program for a PC, 234
  tests, 229
Mentor Graphics, 87
Metal, 63, 64–70, 78–81, 89, 92–95, 116, 117, 122, 123, 245, 247, 249, 250, 252
Micron, 66, 249, 250
Microprocessors, 3, 5, 8, 13, 18, 46, 115, 218
MicroSim, 181, 192, 193
Microsoft, 17, 18
Mobility, 237, 240, 293
Modelling, 137, 138, 159, 181, 201
MOS,
  general, 7–9, 18, 21, 39, 40, 45, 46, 48, 49, 51, 53–55, 65, 72, 74, 75, 237
  characteristics, 53
  device Modelling, 51
  devices as a switch, 54
MOS Technologies, 18
MOSFET, 39, 183, 184, 191, 192, 290, 291
Motorola, 18
MSI, 1, 3
Multiplexer, 118–121, 132, 174, 179, 217
MUX, *see* Multiplexer

—N—

N+ diffusion, 122
NAND, 13, 24–26, 30, 41, 43, 46, 50, 58–60, 63, 72, 75, 79–81, 96, 97, 133, 138–143, 164, 203, 204, 208–211, 219
Nested loop, 138
Netlist, 181, 194, 272
New York Times, 17
NMOS, 8–10, 21, 30, 39, 40, 43, 46, 49, 51–82, 105, 106, 121, 122, 192, 209, 210, 237, 239, 240, 243, 245, 249, 252, 291
NMOS inverter with a depletion load, 56
Noise
  generation, 34
  immunity, 47
NOR, 13, 26, 27, 30–32, 40, 41, 43, 45, 46, 50, 58, 59, 60, 63, 71, 75, 81, 92–98, 133, 138–143, 164, 203, 206, 208, 211, 212
Npn, 21, 39

—O—

Observability, 202
On resistance, 65, 69, 71, 72
Open circuits, 209
Optical disk, 101
OR, 10, 13, 16, 30, 31, 32, 35, 75, 96, 98, 116, 119, 129, 133, 138–143, 153, 154, 164, 178, 203, 206, 211, 222, 224, 225, 227
OrCAD, 87
Overglass, 245, 247
Oxide, 110, 123, 237, 239, 240, 242, 245, 249, 250

—P—

P+ mask, 247
Package, 2, 3, 7, 87, 88, 98, 124, 125, 128, 161, 165, 171, 192, 193, 263, 264–266, 273, 274
Package body, 161
Package declaration, 161
PAL, 16

Parallel plate, 238
Parameter passing, 141
Pass transistors, 121
Path Sensitization, 203
PCB, 14, 65, 66, 117
Pedestrian control, 148–151, 156, 157
Pentium, 8, 63, 113
Permitivity, 237, 240
Permittivity, 237
PET, 18
PGA, 124, 125
Photoresist, 243–247
Piecewise linear transient, 186
Piecewise linear waveform, 185, 186, 287, 296
PLA, 5, 10–12, 16, 116
PLCC, 124, 125
Plice, 122, 133
PMOS, 9, 39, 40, 42, 45, 49, 51, 52, 64, 65, 77–82, 93, 121, 192, 210, 212, 245, 291
Polysilicon, 63–65, 67–70, 78–81, 92–94, 110, 122, 244, 246, 249, 250, 252
Power,
  dissipation, 2, 7, 8, 9, 17, 24, 26, 35, 41, 47, 49, 55, 59, 60, 77, 82, 123, 180, 277
  supply connections, 90
PQFP, 125
Practical FPGAs, 124
Predictability, 202
print_table(), 146
Process,
  general, 16, 173, 262
  statement, 173, 176
process_states(), 146
Program:
  gate capacitance, 241
Programmable switch, 117, 118, 126
Project,
  general, 98, 99, 201
  development, 201
PROM, 11, 103, 116, 130, 131
Protection, 2, 129
PSpice,
  general, 180–200, 263–265, 272–297
  AC analysis, 187, 188
  DC analysis, 180, 181, 188
  device references and model parameters, 183
  units, 181, 282–284, 288–290, 292–294, 297
Pulse,
  transient, 186
  waveform, 185, 186, 287, 296

—Q—

QFP, 125

—R—

RAM, 101, 102, 106, 107, 112, 120, 121, 229
Raytheon, 17, 281
RCA, 17
RED, 64, 82, 83, 85, 99, 148–151, 157
Registers, 3, 5, 217, 218, 220, 223, 226
Requirements,
  analysis, 83
  specification, 83
Resistivity, 73, 191, 245
Resistor, 293, 294
Ripple counter, 87
Rise time, 24, 181, 184–186
ROM, 3, 101, 103–105
Row and column latching, 109
Rows of cells, 91

—S—

Schematic, 71, 72, 87–89, 96, 98, 138, 144, 154, 181, 192, 193, 194, 200, 228, 263, 264–266, 268–273, 277
Scan in, scan out, 218, 219
SDFL, 9
Semi-custom, 12, 13, 15, 16, 63, 115
Sense amplifier, 110–113, 229, 231, 232
Sequential, 173–179, 217, 221, 233, 261, 262
Sequential logic, 144
Shift register latch, 220
Shifted diagonal test, 232
Shockley, 1
Short circuits, 202, 209
show_sequence(), 148
Signature analysis, 222, 226
Silicon, 1, 5, 7, 9, 11, 16, 17, 39, 51, 54, 56, 63, 68, 71, 72, 75, 115, 116, 122,

160, 237, 239, 240, 243, 245
Silicon dioxide, 63–65240, 243–247
SIMM, 108, 109
Simulation, 87, 137, 151, 159, 168, 181, 180, 181, 185, 188, 192, 196, 197, 199, 202, 263, 264, 271–274, 277
Sinusoidal waveform, 185, 287, 296
SiO_2, *see* Silicon dioxide
SISO, *see* Scan in, scan out
Slew rate, 129
Software, 3, 4, 15, 18, 103, 137, 161, 166, 181, 205, 211, 223
Source, 27, 30, 34, 39, 40, 45, 46, 51–56, 60, 63–65, 68, 77, 109, 110, 123, 129, 141, 181, 184, 185, 187, 191, 192, 196, 197, 237, 238, 239, 240, 241, 268, 270, 277, 278, 281, 282, 284–287, 290, 295
Speed, 26, 31, 34, 35, 41, 47, 49, 54, 55, 82, 141, 166, 281
SRAM,
  general, 101, 106, 112, 113, 121–124, 129, 133
  programming, 120, 121
Ssa, 202, 204, 205
SSI, 1, 3
Standard cell, 10–16, 115
State table, 85
Statements, 147
Static, 49, 106
Stick diagram, 64, 70– 72, 75, 78–81, 93, 94
Store and generate method, 221
Structures
  bit fields, 146, 147
Substrate, 16, 63, 77, 79, 109, 123, 191, 192, 237, 243–250, 290–292
Surround disturb patterns, 234
Switched matrix, 125, 127
Switching matrix, 127
Symmetrical arrays, 117

—T—

Temperature, 1, 2, 30, 35, 188, 245, 282, 283, 284, 290, 293, 294
Test patterns, 229
Test vectors, 202, 204, 206, 207, 211, 212, 218
Texas Instruments, 17
THINOX, 243, 246
Threshold voltage, 31, 34, 40, 51, 52, 53–55, 58, 60, 237–240, 245
Timing characteristics, 2, 73, 113
Timing diagram, 145
Traffic lights,
  general, 82, 83, 86, 96, 98, 99, 115, 148, 150, 156, 157, 211, 255
  controller, 88, 89, 98, 99
Transient analysis, 197
Transistor, 1, 3, 7, 8, 9, 11, 12, 17, 21, 22, 24, 26, 27, 30, 31, 39, 40, 42, 45, 48–79, 82, 92, 93, 98, 101, 103, 109, 110, 116, 117, 121–123, 183, 191–193, 209, 210, 212, 237, 239, 240, 244–246, 291–293
Transistors, 1, 3, 5, 8, 9, 11, 12, 14, 15, 16, 17, 21, 24, 26, 31–36, 39, 40, 42–53, 59, 63, 64, 66–69, 72, 75, 77, 79, 81, 82, 101, 103, 106, 115, 116, 120–123, 160, 191, 209, 210, 237, 243, 245
Transmission line, 295
TTL, 8, 9, 10, 21, 22–31, 36–42, 47, 49, 125, 129, 202, 203
TTL inverter, 21, 22, 23, 39
TTL NAND, 24, 25
Twin tub, 77

—U—

ULSI, 3
Ultra-violet light, 103, 116, 123
Unit square, 74
University of California, 181
Unsigned decimal, 147

—V—

VCVS, 284, 285
VHDL,
  general, 137, 159, 161–179, 261
  bit strings, 166
  characters, 165, 170
  constants and variables, 166
  enumeration types, 168
  flow control, 172
  history, 159
  indentifiers, 164

logic levels, 171
numbers, characters and strings, 165
reserved words, 164, 261
scalar types, 167
time and defining units, 167
VHSIC, 159
VLSI, 3, 16, 115, 202

—W—

Wafer, 16, 243, 245, 247
Wait statement, 175
WALK, 82, 83, 99, 148–151, 157
Walking 1's, 233, 236
WALKPAT, 233
Waveform diagram, 83
Western Electric, 17
Windows, 181

—X—

XC2000, 118. 119, 124–129
XC3000, 124, 125, 128–130
Xilinx, 118, 119, 124
XOR, *see* Exclusive–OR

—Y—

YELLOW, 64, 82, 83, 99

—Z—

Zilog, 18

Selected Answers

1.1 LSI
1.3 GaAs
1.4 CMOS
1.5 NMOS/HMOS/TTL
1.6 NMOS
1.7 IIL/CMOS (because of power dissipation)

2.2 8
2.3 40
2.4 14 mA
2.5 9.6 mA
2.6 400 μA
2.7 LOW: V_{in}=0.2V, I_{OL}=–(5–0.7–0.3)/4 mA
2.8 I_1=–1.6 mA, I_2=160 μA, I_3=–12.8 mA
2.9 $\overline{(A.B)+(C.D)}$
2.11 V_{in}=LOW (Q1, Q2 and Q3 off):
V_C=–(5.2–1.4)/2=–1.9V;
V_B=–1.9–0.7=–2.6V;
V_A=–2.6–0.7=–3.3V;
$I_{300\Omega}$=(5.2–3.3)/300=6.3mA;
V_D, and V_O will depend on the load, but with an open-circuit:
V_D=0V.

V_{in}=HIGH (Either Q1, Q2 or Q3 on):
V_C, V_B are the same as LOW input;
The actual voltages and current will depend on the input voltage. Assume that V_{in}=–2.4 V (it must be at least 0.2V above V_B), then:
V_A=–2.4–0.7V=–3.1V
$I_{300\Omega}$=(5.2–3.1)/300=7 mA

V_D, and V_O will depend on the load, but with an open-circuit:
V_D=–7 mA × 300 Ω=–2.1V.

2.12 $\overline{A.(B+C)}$
2.13 Large densities

3.5 NAND, NOR

3.7 UPPER LOGIC BLOCK: (A in parallel with B) in series with (C in parallel with D);
LOWER LOGIC BLOCK: (A in series with B) in parallel with (C in series with D).

4.1 0.0095 pF

4.2 0.0017 pF

4.4 (a) k>=4
(b) k>=8
(c) k>=4
3-input NAND, k>=12
3-input NOR, k>=4

5.1 0.5 mA, 0.11 mA

5.4 (b)

13.1

AND	sa0:	11 {O/P=1}	sa1:	01 {O/P=0}
OR	sa0:	10 {O/P=1}	sa1:	00 {O/P=0}
INV	sa0:	11 {O/P=1}	sa1:	01 {O/P=0}
EX-OR	sa0:	1X {O/P=inverted input }		
	sa1:	0X {O/P=input}		
NAND	sa0:	11 {O/P=0}	sa1:	01 {O/P=1}
NOR	sa0:	10 {O/P=1}	sa1:	00 {O/P=0}

13.4 sa0: 1100, 1101, 1110
sa1: 1000, 1001, 1010

Other answers and tips can be found on the WWW page:

`http://www.eece.napier.ac.uk/~bill_b/mbook.html`